G000165249

WITHDRAWN

ACC. NO. 77650	LOAN TYPE
LOCATION/CLASS. NO. (UP) 509.2 MOR/H	INITIAL Mc
	DATE 29/11/16

Presented to

LATYMER UPPER SCHOOL LIBRARY

by

the library of Professor Richard Perham,
Latymerian and Chairman of Governors,
2005 - 2010.   August 2015

Henry More

HENRY MORE

AND THE SCIENTIFIC REVOLUTION

A. RUPERT HALL

CAMBRIDGE
UNIVERSITY PRESS

Published by the Press Syndicate of the University of Cambridge
The Pitt Building, Trumpington Street, Cambridge CB2 1RP
40 West 20th Street, New York, NY 10011-4211, USA
10 Stamford Road, Oakleigh, Melbourne 3166, Australia

Copyright © A. Rupert Hall 1990

First published 1990 by Blackwell Publishers Oxford
Reissued by Cambridge University Press 1996

Printed in Great Britain by Biddles Ltd, Guildford & King's Lynn

British Library Cataloguing in Publication Data available

Library of Congress Cataloging-in-Publication Data available

ISBN 0 521 56223 6 hardback

Contents

Contents

General Editor's Preface

OUR SOCIETY depends upon science, and yet to many of us what scientists do is a mystery. The sciences are not just collections of facts, but are ordered by theory, which is why Einstein could say that science was a free creation of the human mind. Though it is sometimes presented dispassionately and impersonally, science is a fully human activity, and the personalities of those who practise it are important in its progress, and often interesting to us. Looking at the lives of scientists is a way of bringing science to life.

Those scientists who appear in this series will be chosen for their eminence, but the aim of their biographers is to place them in their context. The books will be long enough for authors to write about the times as well as the life of their subjects. Science has not long been a profession, and for many eminent practitioners of the past it was very much a part-time activity; their *Lives* will therefore show them practising medicine or law, fighting wars, looking after estates or parishes, and not simply focus upon their hours in the laboratory. How somebody earned a living, made a career, got on with family and friends is an essential part of a biography: though in this series it is the subjects' commitment to science that has got them in, and must be always at the back of the biographer's mind.

Henry More's name will not be familiar to most of those who practise science today. There is no law, chemical reaction or disease named after him; no laboratory or scientific society bears his name; and indeed it would be curious to call him a scientist at all. The word was coined in the 1830s by analogy with 'artist'; by then there were sufficient numbers of people active in the natural sciences to need a term to describe themselves. They were no longer philosophers or savants with special interests in nature, but a self-conscious group engaged in a common enterprise. Even in the nineteenth century, the term took some years to catch on, and 'natural philosopher' or 'man of science' remained, with 'naturalist', popular especially with those reluctant to see their common culture divided into humanistic and scientific halves.

In the middle of the seventeenth century, there was the 'New Philosophy' of Bacon, Galileo and Descartes, challenging the beliefs inherited from ancient Greece; this involved an emphasis upon answerable and empirical questions, and led to the beginning of modern science. But More did not devote himself to fact collecting, experiment or mathematics; he became one of the most celebrated philosophers of his day, corresponding with Descartes, and playing a very important part in transforming Cambridge from a stronghold of Puritan orthodoxy in the early part of the century into a more liberal place where mathematics could flourish in the Restoration period.

More was one of the group called the Cambridge Platonists, who saw Plato as a kind of prophet of Christianity, and he was also a believer in atomism. The atomic theory in the ancient world and in the Renaissance had been associated with irreligion; the gods of Epicurus and Lucretius watched the collisions of the atoms, but had not created them and could not intervene in the world. More, in his poem *Democritus Platonissans*, tried to combine the insights of Plato and those of the atomists, arguing for created particles and for the possibility of divine interventions. He believed that the atoms, possessing only the primary qualities of the new philosophy (motion, weight and shape), gave rise by their various arrangements to the secondary qualities, such as taste, colour and smell.

For More and his associates, as for Plato, God had worked through Nature, a kind of demiurge, in the creation, rather than doing everything himself, and the result might therefore be less than perfect. Robert Boyle, among others, disagreed with this, refusing to see an intermediary between God and the world as he refused to see one between God and humanity — he was a good Protestant. But More has an important place in making atomism respectable in Britain, and in thus laying foundations for the work of Boyle and of Newton.

More was a prominent figure in Cambridge when Newton came up as an undergraduate, and recent work on Newton has pointed to the importance of the philosophical background to Newton's thinking. Rupert Hall, in this fascinating life, has come to More after editing Newton's later correspondence and various published works, and nobody could be better qualified to guide us through the new philosophy and the origins of modern science, intellectual and social. More was a very important figure in his own day, not a great original thinker like Descartes or Newton, but a synthesizer bringing together new and old thoughts into a new pattern. Against the background of thinkers like him we can assess the originality of Newton properly, and without understanding men like More we find that the greatest thinkers become inexplicable. This life will provide a splendid route into the middle of the seventeenth century, and its Scientific Revolution.

David Knight
University of Durham

Preface

Henry More (1614–87), Fellow of Christ's College, Cambridge, was the greatest of English metaphysical theologians and the most perplexing, perhaps also the most distinguished, member of the group of seventeenth-century divines known as the Cambridge Platonists. He is in several respects a paradoxical figure. An admirer of Galileo, Descartes and Boyle, he rejected their detailed applications of the 'mechanical philosophy' to the explanation of natural phenomena. He was an experimenter, yet also a cabalist. He applauded the rising scientific movement of his day and became a Fellow of the Royal Society, yet maintained that the forces of Nature were spiritual. Isaac Newton approved More's atomist philosophy but rejected his 'Spirit of Nature'.

Because Henry More was an important figure in the history of my own College, his name has been familiar to me for fifty years; later I sat often under the sad portraits of his friends, Finch and Baines. Before the war Sir John Plumb assured me that Henry More had seen the devil flying round the tower of Great St Mary's, clad in leather breeches. I have not found such a tale in his writings, but there are others there nearly as strange. Copies of More's books and an important collection of his manuscripts are preserved in the College Library.

More was a prolific and varied author in both English and Latin. As several of his books were reprinted in his lifetime and since, he has been widely read. But if More is often seen on the stage of history, he has rarely come under the spotlight; if he often figures in articles and monographs upon intellectual history, few books have been devoted wholly to a consideration of his writings. Yet his work touches on many points of perennial interest, not least his relation to the scientific movement of the seventeenth century, which he both aided and opposed. The young More popularized in English verse the work of Galileo and Descartes; the mature More ignored the former and rebelled against the latter. In final judgement More may be found an opponent of the rising scientific spirit, even if his ideas did contribute positively to Newton's concepts of space and time.

My book is concerned only with the scientific aspects of More's thought, and his relationship to Newton in particular. I have excluded any evaluation of More as poet and theologian. In his biography I have emphasized his scientific relationships, and in writing of his philosophy, his philosophy of Nature. I have dealt lightly with More as a figure in the history of witchcraft and spiritualism in England, and upon his involvements with cabalists, Rosicrucians and alchemists. His detestation of ranting religious enthusiasms seems to me to belong to a treatment of More as a religious writer. It is perhaps a more serious (though deliberate) omission that, though I have tried to put together an accessible account of More in the contexts of Renaissance and Cambridge Platonism — for some grasp of these contexts is essential to an understanding of his scientific writings — I have attempted no full assessment of More as a Platonist; that would require a different book.

The aspects of More's writings least examined by me have been fully treated by Dr R. W. B. Crocker in his Oxford D. Phil. thesis 'An Intellectual Biography of Henry More (1614–87)', presented in 1986. He has studied More as a mystic and a man obsessed with the spiritual element in the world. Dr Crocker's is undoubtedly the most thorough analysis of More to date, but More's relation to or possible influence upon the scientific movement of his time, central to my interest, is tangential to his. I read his thesis after my own book was written and was happy to discern no obvious conflict between us. I must have some recollection also of a Cambridge MA thesis by Miss P. M. L. Moir on 'The Natural Philosophy of Henry More', but it is twenty-two years since I examined it with Dr Gerd Buchdahl. Dr Crocker has compiled elaborate bibliographical materials which will be of great use to future scholars.

Preface

Part I of this book is introductory to the more detailed study of More and the Scientific Revolution in Part II. I have not examined every one of More's numerous writings with great care because many are irrelevant to my purpose. I am much indebted to the work of other writers, above all to the *Conway Letters* of Marjorie Hope Nicolson (1930), an invaluable source. Appendix I lists More's chief philosophical writings. The Bibliography lists many published studies relevant to my topic.

For assistance in preparing this book I am indebted to the Librarian of Christ's College, Cambridge, and Mrs Courtney; to Drs Alan Gabbey and John Henry, who have explored More before me; and to Dr Sarah Hutton who kindly invited my participation in a symposium on the tercentenary of More's death. Particularly I am indebted to Sir Hans Kornberg FRS, Master of Christ's College, whose invitation to deliver a commemorative lecture in the College induced me to prepare this introductory study.

A. Rupert Hall
Tackley, Oxford

Cambridge! Cambridge! What a monstrous mother art thou! I never thought the *same womb* could labour with *Moores* and *Christians*.

Thomas Vaughan, *The Man-Mouse*, 1650

Part I

Platonic and
Personal Background

1

Introduction

O F ALL THE historically distinguished products of Christ's College, Cambridge, Henry More is the most enigmatic. Choosing a first-eleven team of Christ's men, scientists would pick Charles Darwin as Captain, humanists John Milton. Along with William Paley of the *Evidences*, William Lee of the stocking-frame (as near a progenitor of the Industrial Revolution as one can find), More's friend Ralph Cudworth (fourteenth Master of the College) and various public men of the present century, More would certainly deserve his place. He would also be, almost certainly, the only member of such a team to have passed his whole active life in Christ's College.

It would be a mistake, however, to imagine that among the great More is as obscure as he is enigmatic. He figures in every history of philosophy and in most general histories that touch at all on the intellectual life of his time. His writings were of no small interest to greater philosophers such as Locke, Leibniz and Spinoza. He appears in a recent anthology of verse in the manner of Edmund Spenser and four at least of his books have been reprinted in modern facsimiles. Every year fresh articles and books on More are printed, though few seem to throw a clearer light upon his life and mentality. More's thought is far-ranging, obscure, hastily and carelessly expressed in many instances and not a little inconsistent. It reflected a very ancient tradition that More sought ineffectually to modernize, a tradition that is intrinsically esoteric and indeed nebulous, one wholly alien to contemporary ways of thinking about the universe, humanity and religion. More was writing in harmony with many themes of his own age, but they are themes that have long ceased to resonate in our own ears.

Henry More is the only metaphysician of importance in the history of Christ's College, arguably the most important in the history of a pragmatic University. The other 'glass-worthy' thinkers (to repeat Peile's allusion to the College Hall) were direct and lucid (1900: 32).[1] Charles Darwin, a man of great psychological complexity, was open, clear, matter-of-fact in thought and expression. The creed of Paley, one of the great natural theologians, was a simple one. Cudworth's *True Intellectual System of the Universe* finds only specialist readers now, but this is rather because of the weight of his learning than because his thought was subtle. John Milton is a greater figure in our literature than either More or Cudworth, and his writings impose their own perplexities. However, I submit that at least as a prose author Milton was plain and hard-hitting to the point of brutality, and I am not sure that even *Paradise Lost* can be said to involve difficult metaphysical notions.

Henry More is that strange and rather sad figure, a major but undisciplined talent in a minor intellectual movement that was destined to be overridden by history. A man of enormous literary energy and output, intellectually curious, widely read, highly esteemed by his contemporaries and since (for his books were continually reprinted for centuries after his death), yet fated to enjoy no time of triumph. There was never a moment when all the world talked of the Cambridge Platonists, as it talked of *Paradise Lost*, the *Origin of Species* or the *Evidences* of Christianity. All these books remain to this day more intelligible than those of Henry More, partly because More wrote in a style that we can follow even less well than that of John Donne and the metaphysical poets, to whom More was intellectually akin. It is an effort for even the devotee to peruse *Divine Dialogues* between characters bearing such names as Hylobares and Bathynous, and More wrote in Latin, as well as English verse and prose. He was not only a metaphysician but above all things a Christian theologian and philosopher, the divine and pious Dr More, the Angel of Christ's. A joyous defence of Christianity by the shield of Platonic philosophy was the centre of his youthful being. More was a poetically religious metaphysician whose ideas of God and Nature were glowing, romantic, mystical and often barely intelligible.

INTRODUCTION

To characterize More's intellectual attainments one might say that he was above all occupied with the insubstantial world of spirits. In this were three principal constituents: the Spirit of God, everywhere ruling the universe; the Spirit of Nature, akin to what the ancients had called the *anima mundi*, the soul of the world, or Nature personified; and the Spirit or soul of Humanity, equally immortal, and capable of being active for good or ill in this material world (as More believed) even in immaterial form, as a ghost or apparition. Samuel Johnson praised More for his penetration into this mysterious realm, whither no one has trodden with greater assurance. To More the spirit world sanctified by Christianity was the equivalent of the Ideal world of Plato: that is, the Spirit of God is the Idea of God; the Spirit of Nature is the Idea of all that exists in the universe, and the Spirit of Humanity is the Idea of humanity. The material universe and the human body are no more than contingent, existential realizations of the Ideas of Nature and Humanity that result from God's choosing to create material existence and permit it to run its course.

But Henry More was not only a writer whose books are more remarkable for imaginative prolixity than for disciplined thinking of the German type. He was also a Christian theologian, though again of a somewhat impressionistic kind. The first object of his life was to lead his fellow men to Heaven. He felt strongly the truth, power and beauty of his religion; he tried by a variety of methods – metaphysical, ethical, natural-philosophical – to persuade his readers of the same. Again and again he comes back to the Bible as the prime source of human knowledge and the only sure guide to the salvation of the soul, as all the Cambridge Platonists do. Like Isaac Newton (of whose opinions he did not wholly approve) he devoted much labour to unfolding the meaning of its prophecies. Like his fellow Platonists too, More's ideal was a simple religious sincerity; in Cudworth's words, 'he is the best Christian whose heart beats with the truest pulse towards Heaven, not he whose head spinneth out the finest metaphysical cobwebs.'[2] For Christian Platonists from Steuco onwards Christianity was not a tight-rope of theological doctrine that the true believer must walk with perfect poise, but a matter of life, love and conduct. All seventeenth-century philosophers including Spinoza but excepting Thomas Hobbes take the perfection and beneficence of God as the foundation of every branch of human enquiry, without the assumption of which reason itself is vain. So insists More.

As a literary scholar, that is, in the basic skills of his trade, More is nowhere near the class of Isaac Casaubon or J. J. Scaliger. He had a sound but unremarkable knowledge of authors in Greek and Latin, especially the neo-Platonists and the Hermeticists made known by Marsilio Ficino, and a working knowledge of Hebrew. He was not concerned about the detailed study of texts and techniques of philology. He was a little of a geometer and knew something of astronomy, which figures a good deal in his philosophical poems; diagrams and technical discussions are to be found here and there in his writings. He was not much interested in the academic world outside Cambridge and the bulk of his extant correspondence – which is mostly with close friends – is relatively small. More liked to think that he had a competent knowledge of natural philosophy (as well as of such other specialist branches as ethics), taking it upon himself to dispute on matters of mechanics, astronomy, hydrostatics and pneumatics with such eminences as Descartes and Robert Boyle, among others. Unlike some critics of Galileo, Descartes and Newton, More did not stand as an apologist for some older natural-philosophical tradition, such as that of Aristotle; on the contrary, More was often (up to a point) in sympathy with the 'moderns', particularly in his earlier years. But he was far more warmly devoted to the defence of Christianity from any possible danger that might arise from the speculations of natural philosophers.

As regards his own competence to dispute with them on their own specialist ground, More was simply mistaken. Natural philosophy was becoming increasingly mathematical and experimental in the seventeenth century; More possessed neither mathematical nor experimental abilities. What is more serious, his understanding of the methods and purposes of natural philosophy seems to have diminished during the course of his life, precisely while its fruits were becoming more rich. The endeavour of natural philosophy since the Greeks had been to identify the causes of things within the normal composition of Nature itself (strange as some of these agencies may seem to us) and to avoid the invocation of causes lying outside the composition of Nature. Gods, spirits of grove and stream, were thus ejected as causes of natural phenomena. This endeavour we may label rationalist. The constructive trend of the Scientific Revolution was towards a further narrowing of the definition of what might count as within the ordinary composition of Nature, outside of which a supposed cause would be a miracle. The Spirit of Nature postulated by Henry More could not figure within the composition of Nature as natural philosophers now conceived it; nor indeed would it have been admitted within the ancient mathematical tradition of Euclid and Hipparchos, Archimedes and Ptolemy. In fact, though he seems to have been blissfully unaware of the fact, More does not object against Boyle alone in hydrostatics, but Archimedes also.

Thus there was a conservative retrogression in Henry More's life. The poet who helped set the ideas of Copernicus, Galileo and Descartes before the English in the 1640s became, twenty years later, the crabby and obtuse opponent of the new mechanics and pneumatics. Of course, there were good theological motives for More's opposition to what delighted the Royal Society – though, I hasten to add, he had no quarrel with that body of which he was himself an inactive Fellow. But it is important to note that More's opposition was not theologically *argued*. Opponents of evolution in the nineteenth century declared: this notion denies the teaching of Scripture, therefore it is false. More argued in the opposite sense; he said: this notion of the natural philosophers (for example, the notion that air is elastic, possessed of an inherent springiness) can be proved to be false, for it is absurd, therefore its potential danger to religion is nullified. (The danger to religion arose from the attribution to mere brute matter of an intrinsic power of action, such as elasticity, much as Newton later refused to allow to mere brute matter the active power of gravitation.) Though More's objective was the defence of religion, his argument is about true or false in natural philosophy. But because More understood neither the concepts nor the experiments of those whom he criticized, his attempts failed, as Boyle, Hooke and others forcefully pointed out.

Although More had in earlier days praised Descartes as the 'sublime Mechanick' because (as More then believed) Descartes had given the best attainable explanation of the workings of Nature without invoking the power of Spirit (and therefore by its defects showed just how important Spirit must be), in middle life More passed to the extreme view that no ground was to be left for the natural philosopher to occupy with his rationalist explanations. Every kind of activity in Nature was to be attributed to Spirit, just as every human activity was to be attributed to the volition of the soul. More's pupil, friend and patron, Lady Anne Conway, went far beyond More in denying the real existence of matter altogether: matter, she believed, was simply an inspissated manifestation of Spirit. More was reluctant to push to obvious conclusions along the road he had taken. But it seems that in the end he was saying that the natural philosophy of Descartes or of his own colleagues in the Royal Society could explain nothing, not even the motions of two billiard balls impacting one upon the other. Its role could only be to describe events in Nature and look for patterns in them — much as the positivists believed centuries later. The deeper truths or theories, the laws and operations of the Spirit of Nature, were not to be found out by mathematics and experiment. There is, of course, good Platonic precedent for this way of thinking.

Two interesting biographical questions therefore present themselves. How did More become a Platonist? We know that his devotion to Plato was born during his early years at Christ's College out of a revulsion from the religion of Calvin and the philosophy of the Schools — the second-hand philosophy of Aristotle — but how it was nourished, where the books came from, we do not know. One can speculatively link More's intellectual development with Joseph Mede and other Christ's men of an older generation, but we have few solid facts other than that there was an interest in Plato and the neo-Platonists at Christ's, like that better documented for the sister College, Emmanuel. Then one may again ask, how was More introduced to mathematics and the reading of Galileo, followed (after 1644) by that of Descartes? This question is made more cogent by the pessimism of Christopher Hill and others about the state of learning in the English universities. However, the fact seems to be that Hill's utter gloom is needless. Christ's men senior to More, like William Perkins and Samuel Ward (later Master of Sidney Sussex), Robert Gell (Fellow, 1623–38) and Nathaniel Tovey (Fellow, 1621–45), demonstrably were active in mathematical and physical subjects. There is evidence that Mede in particular encouraged undergraduates to pursue them eagerly, and that undergraduates did indeed attend lectures on mathematics and philosophy, as they were supposed to do (Feingold 1984: 60; 62; 81; 96; 110). All this, however, though informative in a general way, still leaves us in the dark concerning Henry More, the individual. The only other consideration I can offer here is that Henry Burrell (d. 1627) left money to Christ's College for the purchase of mathematical books; many that were bought from this fund, among them books by Galileo and Kepler, still remain in the Library (Peile 1910: 267).[3] It may be that these were accessible to More.

At all events, we cannot simply isolate More from the Scientific Revolution of the seventeenth century and resign him to the historians of letters, ethics, cabalism, metaphysics and theology. He not only possessed (as he supposed) overwhelming theological reasons for speaking out against the predominant scientific trend of his day, which we now call the mechanical philosophy of Nature, but strong natural-philosophical reasons also, founded upon the experimental evidence, for so doing. It was his purpose to turn natural philosophy away from the direction it had taken ever since the late sixteenth century, and to redirect it towards older, sounder, Platonic modes of thought. In this he did not succeed. But the intellectual difficulties in the mechanical philosophy as More knew it from Descartes, and even from the Royal Society, did not go unmarked by others also, and it may be that More had some effect in modifying the mechanical philosophy, and in consequence the concepts of physics, in ways not foreseen by him. To this point we shall return.

To leave an introduction to the thought of Henry More at this point would perhaps be unfair to him, for More's writings against the English philosophers of mechanistic Nature constitute only a tiny fraction of his entire output, and the destruction of the mechanical philosophy was not his chief objective, not even his chief negative objective. The mechanical philosophy was only one element in the atheistic materialism that was More's principal target. Moreover, it would be unjust also to dwell only on the negative, critical side of More's writings. His positive proclamations of the existence of God and the immortality of human souls, his lengthy expositions of the role of spirits in the Universe, both beneficent and devilish, his exaltations of the long and divinely inspired idealist stream in the evolution of human wisdom of which true Christianity was the supreme expression, such elements as these form the major part of his vast output. To evaluate More the philosopher 'in the round' would require a book of great scope far exceeding the confines of my more limited investigation. It seems appropriate, then, to conclude this introduction with the independent opinion of an American scholar whose perspective on More is wider than my own:

> the Cambridge Platonists, and particularly More and Cudworth, were extremely influential in their generation and beyond it. Their place in the history of English philosophy, in particular of English idealism, is an established one; the more we know of English latitudinarianism, of English

Cartesianism, of English materialism in the seventeenth century, the more we realize how deeply involved the Cambridge Platonists were in the fabric of seventeenth-century social and intellectual life. (Colie 1957: 6)

I do not believe that many historians of philosophy would say less.

2

Platonism

THERE ARE hundreds of books on Platonism, and as many definitions of it. There are at least as many more on neo-Platonism, the mystical philosophy created centuries after the Greek philosopher's death by a process of selection and re-creation. For some, Platonism is the power of mind to understand and indeed control the universe of matter and, as the philosophy that inspired the mathematical interpretation of Nature, Platonism, on this view, has furnished the plan for the creation of modern science. The philosopher and historian Alexandre Koyré taught that it was the Platonism of Galileo that enabled him to reform physics (1968: 16–43). To others Platonism, with its emphasis on the intuitive and the *a priori*, has seemed the antithesis of true science which can rise only from painstaking investigation of the natural world: numbers must derive from measurements, not imagination. The historian George Sarton found 'hot air' in Platonism, too often purveying 'magic and non-sense'. To him, the 'history of Platonism is the history of a long series of ambiguities, misunderstandings, and prevarications' (1953: 426; 436n; 451).

In the most obvious sense, a Platonist might be defined as one who delights in the writings of the Master; there have been many such through the ages, though with little enough agreement between them on the meanings and marvels of the Platonic dialogues. The great editors and textual students of Plato have not all been Platonists in the accepted sense of the term. Most generally, the Platonist is an idealist philosopher, one who believes that the essential truth of material things lies either in some soul or spirit informing them, or in their invisible architecture, its plan, pattern and structure. This truth is something not seen with the eye of the beholder who contemplates the outward form of flea or galaxy; it is internal, unrealizable, to be seen only by the eye of the mind. We may say that the skeleton of an animal defines its essence, but we do not intend the dry bones wired together by an anatomist. It is the skeleton live, articulated and moving, invisible beneath skin and flesh, and not the skeleton of one poor dead creature but the generalized skeleton of a set and pattern of creatures, as, for instance, of the family *Felis*. The composite visible object, the wired skeleton, merges into the *idea* of structure and form, and beyond that into concepts of taxonomical order, adaptation and evolution. The Platonist believes that the commonplace antithesis between fact and theory is altogether false. There are no such hard simple things as facts and measurements of whose truth we become certain by using our eyes and fingers (like the doubting Apostle Thomas!) and such aids as microscopes or Geiger counters. Nor on the other hand are there such things as theories to be regarded as belonging to a sphere of thought and explanation quite distinct from the positive world of facts. Platonists do not believe in the necessity to modify theory to suit new ratiocinations, or to accommodate newly discovered facts. On the contrary, it is in their view the world of theories or ideas that forms our world of facts; it is by means of our theories that we satisfy ourselves of the existence of facts, without them we would have no means of distinguishing fact from fancy, truth from error. Is *this* a quark or a snark? Only by knowing what world of thought it is that we inhabit can we mark the distinction.

There have been Platonist biologists, such as the philosopher Goethe who traced the modified repetition of archetypal forms everywhere in Nature. But the paradigm of the Platonist is the geometer, because Plato believed geometry, the fundamental knowledge of form, to be the key to all understanding of Nature, and also because in the *Timaeus* Plato set out, in a very elementary sketch, a model of how form may be the origin of the perceptible differences between the qualities of things, such as hard and soft, hot and cold. So the Platonist not only believes that the invisible provides the explanation of the visible, but also that the quantitative is the explanation of the qualitative. However far one proceeds in analysing the material structure of things, from the skeleton back to the component molecules of bone, from the chemical molecules back to electrons and protons, and then again to whatever further level of particle structure the physicists may discover, still one does not reach the limit or even the basis of understanding material things. The limit is beyond the picturable, in the world of mathematical forms which have no representation in three or perhaps n dimensions. For the true Platonist the transition from physical form to mathematical form is the crossing of a boundary of principle, the boundary between the material world explored by the senses and the world of ideas explored only by the voyaging mind, a world of which only conceptual understanding is possible. The world of ideas is the true reality of which material existence is no more than a crude simulacrum. Therefore conceptual understanding, the higher kind, cannot be derived from the investigation of material things – as the Christian Platonist Henry More would have said, it can come only from God.

An idealist philosophy in which the brute strength of matter and the brute thrust of force resolve into the impalpability of mathematical equations seems less bizarre in our time than it did to some of our nearer ancestors. If we have given up hope of Eddington's genesis equation, we now look to mathematical physics for an account of the origin of all things in the 'big bang'. We recognize the merit of Johannes Kepler's recollection of the Platonist microcosm–macrocosm analogy when he exclaimed, inspired by the elegant geometrical symmetry of the snowflake, 'From this almost nothing I have found the all-embracing universe itself!' (1966: 39) We may not now share the confidence of Sir Joshua Reynolds – repeating the message of Luca Paccioli and Albrecht Durer – that geometry provides the sole base of beauty. On the other hand, the action of Reynolds's friend, Samuel Johnson, in kicking a stone to refute the idealism of Bishop Berkeley strikes us as merely obtuse. If kicking the stone would refute Berkeley it would also refute not Plato alone but Newton and every philosopher who has recognized that our sensory apprehensions of the natural world, unrefined by intellect, produce illusion; that we perceive reality only (as Plato expressed it) in the form of shadows thrown by firelight. Of course there have been philosophers and scientists who have found all the reality they needed (or regarded as attainable) at the accessible, picturable level and have doubted the meaning of any conceptual account of Nature claiming greater profundity. For Aristotle the qualitative transformations of matter represented the limits of knowledge. Alternatively, in the long atomic tradition down – one might almost say – to 1925, that limit was found in the kinematics of hard bodies. In the positivist tradition of the nineteenth century, *Naturwissenschaft* was the totality of descriptions and pointer-readings; all else was uncertain, mutable speculation. Even today we recognize that it is where science is most conjectural that it is also most Platonist.

Furthermore, the Platonism of modern science – if such there be – continues only a minute fraction of the historic Platonic tradition. Long before the end of the ancient world, Platonism had come to signify far more, and far more that was dubious, than the search for a world of geometrical ideas underlying the world of tangible reality. In philosophy – a topic having more appeal to the Greek than to the Roman mind – the last centuries of the Roman Empire were marked by the evolution of a 'neo-Platonic' system beginning with Plotinos in the third century AD and ending with Proclos in the late fifth. Plotinos taught in Rome itself; the later neo-Platonists flourished in Syria and Greece, where the school of philosophers at Athens, claiming descent from Plato's Academy, was at last shut down by the Emperor Justinian in 529. Several of these late masters were, in the words of Edward Gibbon, 'men of profound thought and intense application; but, by mistaking the true object of philosophy, their labours contributed much less to improve than to corrupt the human understanding ... Consuming their reason in ... deep but unsubstantial meditations, their minds were exposed to illusions of fancy' (1910, I: 382). Far the most illustrious of the converts to this reinvigorated philosophy and religion of pagan neo-Platonism was the Emperor Julian, the Apostate. After his downfall, the intellectual history of Byzantium, the eastern Christian state established by Constantine the Great, was to be characterized by Christian fervour and a revival of Aristotle's philosophy, but the first great commentator on Aristotle's writings, Simplicios, had been trained in philosophy by the neo-Platonists.

Of all Plato's works only one, the *Timaeus*, can be regarded as containing important precepts or opinions about the interpretation of Nature, and some scholars (George Sarton among them) have treated Plato's *a priorist* arguments and contempt for observation as justifying their labelling him as an enemy of natural science or at least as a false prophet. That successors to Plato who were even more imaginative and more mystical than their master should be more painstaking and factual than he in their investigation of Nature is not to be expected. The Greek neo-Platonists were much concerned with their understanding of the 'One' and the 'Good' (perhaps to be identified) and with the efforts that philosophers should make to raise themselves to a higher level of being. Theirs was a mystical, moral and religious philosophy through which philosophers looked inwardly towards themselves, not outwardly at the surrounding universe: mere matter was the lowest form of being. The Platonic Idea – the antithesis of matter – was in effect personalized to become a divine omnipotence though not an omnipotent Creator in the Christian sense. The Platonists believed in a multitude of hierarchies of existence though they also emphasized unity as the counterpart to universality. So in the universe as in the individual everything was interdependent: a change in one part must affect the whole and, because the universe is hierarchical, the higher parts (the heavens) must strongly influence the lower (the Earth). The Sun, as the Platonists rightly saw, rules all life on Earth which (they believed) was also subject to the influence of the Moon and the Planets. Here was authority for astrology. The same line of thought furnished the grounds for the supposed correlation or inter-relation more generally between the *macrocosm* (the universe) and the *microcosm* (the Earth, or more specifically the human body). The neo-Platonists were fond of echoing Plato's dictum concerning the contemplative mathematical sciences (*Republic*, Book 7) that 'the soul through these disciplines has an organ purified and enlightened, which is blinded and buried by studies of another kind, an organ better worth saving than ten thousand eyes, since truth becomes visible through this alone.' Such words as *truth* and *knowledge* commonly in these traditions carried a moral meaning: to Archytas the Pythagorean is attributed the assurance that scientific knowledge restrains people from injurious conduct, for they realize that the guilty party must be discovered.

In composing the world of experience from geometrical elements, Plato had started from the line, which is unity, then argued that a surface is a compound of lines, and similarly that a volume (of which the tetrahedron is the simplest) is a compound of surfaces. Volumes, three-dimensional bodies, constitute the world of experience. In *Timaeus* Plato marries solid geometry to the theory of the four elements by associating the pyramid with fire, the cube with earth, the octahedron with air and the icosahedron with water. What goes with the regular solid with the largest number of sides, the dodecahedron? It represents the whole universe. Plato, no atomist, did not mean that the fundamental particles of each element were of these shapes. Precisely how his ideas should be interpreted is dubious, and (according to Sarton): 'It is typical of the harm done by the *Timaios* that many scholars have racked their poor brains and perhaps been driven to distraction and insanity by the puzzle offered to them, in such solemn terms, by the divine Plato' (1953: 436).

About the time of Constantine the neo-Platonist Iamblichos substituted for the essentially geometrical metaphysical notions of Plato an arithmetical numerology deriving from the parallel Pythagorean school. Such numerology as his, and number-theory in the technical sense of the mathematician, have a common origin in the Pythagorean recognition of number forms and patterns: besides the familiar series of squares, cubes and higher powers of the integers they knew such series as those of the perfect numbers (an integer which is equal to the sum of its divisors, including unity, e.g. 6, 28, 496, 8128, . . .) and the triangular numbers (successive sums of the series of natural numbers 1, 2, 3, 4, . . ., e.g., 1, 3, 6, 10, 15 . . .). Some of the general rules for forming such series were stated by Euclid. Iamblichos was the first to record 'amicable numbers' (where each equals the sum of all the divisors of the other, instancing 220, 284– no further pairs were discovered before the seventeenth century).

The peculiar feature of neo-Platonic numerology is the attachment of ethical significance to particular integers and to the numbers that appear in patterns, following the example of the Pythagoreans who had sworn a holy oath upon the triangular number ten, or the *tectractys*. Words and names could be 'encoded' by the summation of the numerical value assigned to each letter (e.g. $a = 1$, $b = 2$) in order to obtain their moral or prophetic significance.[1] So, according to Iamblichos, Pythagoras had ordained that the number of libations offered to the Gods should be three, that sacrifice should always be made to Venus on the sixth day while sacrifice to Hercules should be made on the eighth day of the month, because he was born in the seventh month (Taylor 1818: 111). Number, that is, intelligible knowledge of the cosmos, the Pythagoreans had called 'the wisest thing'. Theirs, too, was an esoteric sect whose members were sworn to secrecy concerning their beliefs and practices, among which the theorems of geometry were included. Hippasus, a Pythagorean who divulged the method of forming a sphere from twelve pentagons, perished in the sea because of his impiety; the same story is told of the divulgation of the great Pythagorean mystery that the diagonal of a square is incommensurable with its sides.

One may imagine that for the neo-Platonists geometry was not merely the inscribing of lines, nor arithmetic merely counting. Mathematics possessed an esoteric significance, the philosopher distinguishing intellectual numbers from the ordinary numbers of commerce. The former enjoyed a demiurgic or fabricative power, imitated by the ordinary numerals. The Pythagorean Eurytos said that:

> a particular number is the 'boundary' of *this* plant, and again another number of *this* animal, just as six is the boundary of a triangle, nine of a square, and eight of a cube. As the musician harmonizes his lyre through mathematical numbers, so Nature through her own natural numbers orderly arranges and moderates her productions. (Taylor 1818: 316)

The participation of the heavens in numbers was evident to all, but number was equally important in generation, in ensuring that each birth followed as it should the species of its parents. 'In short, physical numbers are material forms divided about the subject which receives them. But material powers are the sources of connection and modification to bodies. For form is one thing, and the power proceeding from it another' (1818: 317). Knowledge of these things, and control of the natural powers that such knowledge confers, is magic.

It is possible that the later neo-Platonists of antiquity drew from the school of Pythagoras, rather than from their avowed master, the strong leaning towards magic that characterized their sect. Gibbon remarked upon this strange transformation:

> [The neo-Platonists] flattered themselves that they possessed the secret of disengaging the soul from its corporeal prison; claimed a familiar intercourse with daemons and spirits; and by a very singular revolution, converted the study of philosophy into that of magic. The ancient sages had derided the popular superstition; after disguising its extravagance under the thin pretence of allegory, the disciples of Plotinos and Porphyry became its most zealous defenders. (1910, I: 382–3; cf. II: 366)

Certainly the Pythagorean religion as depicted by Iamblichos had been full of seemingly irrational prohibitions and injunctions. Thus, the right foot must always be inserted into its shoe first, but the left foot placed first into a footbath. A man may be helped to pick up his burden, but not to lay it down. A white cock may not be sacrificed, for it is a suppliant and sacred to the Moon. A man must enter a temple from the right side and leave it from the left: 'For the right hand is the principle of what is called the odd number, and is divine; but the left hand is a symbol of the even number and of that which is dissolved [that which perishes]'. The true Pythagoreans were vegetarians, but they were forbidden to use the mallow as food because it is the first messenger and signal of the sympathy between celestial and terrestrial natures (an observation of phototropism?). As is well known, Pythagoras strictly forbad the eating of beans and of any other food likely to cause flatulence: the Pythagorean Timycha, wife of Myllias the Croronian, bit off her own tongue so that even under torture she could not reveal to the tyrant Dionysios of Syracuse the reason why members of their sect refrained from eating beans or touching them in any way. To Pythagoras himself (whose sacred name his devotees did not pronounce) supernatural powers were attributed, such as the ability to hear the music of the spheres and to bring about medical cures by applying the principles of harmony. The neo-Platonists believed that Pythagoras had acquired magical powers during his long travels among the eastern sages: the magi of Babylon and the priests of Egypt. Magic was the highest part of the priestly function and was employed to make the temple lamps light up or its statues speak. It was also used to work cures, temples being the chief resorts of sick people.

Proclos, the last and most mystical of the ancient neo-Platonists, explained how the source of magic lies in the alliance and sympathy between natural things, so that all things subsist in all. The ancient priests, he wrote, 'recognized things supreme in such as are subordinate, and the subordinate in the supreme: in the celestial regions, terrene properties subsisting in a causal and celestial manner, and in Earth celestial properties but in a terrene condition' (Taylor 1818: 299). The terrene quality of the Sun is manifest in the circular dance of the heliotrope, and the sunflower following its passage across the sky, while conversely animals, plants and stones possess an intellectual life in the celestial regions. Some 'solarian' animals, such as lions and cocks, participate according to their natures with the divine quality of the Sun.

Hence [writes Proclos] the authors of the ancient priesthood discovered from things apparent the worship of superior powers, while they mingled some things and purified others. They mingled many things together because they saw that some simple substances possessed a divine property though not, taken singly, enough to call down that particular power of which they are participants. By the mixing of many things together they attracted upon us a supernal influx; and by the composition of one thing from many, they produced an assimilation to that One which is above many; and composed statues from the mixture of various substances conspiring in sympathy and consent. (1818: 303).

From the last phrases it is evident that Proclos was speaking of temple magic. Plotinos gave a fuller account of the thought behind the practice:

I think ... that those ancient sages, who sought to secure the presence of divine beings by the erection of shrines and statues, showed insight into the nature of the All [God]; they perceived that though [the] Soul [of the World] is everywhere tractable, its presence will be secured all the more readily when an appropriate receptacle is elaborated, a place especially capable of receiving some portion or phase of it, something reproducing it and serving like a mirror to catch an image of it. (Yates 1964: 64)

The statue is thus a 'wireless receiver' as it were of an ubiquitous message, to which it is attuned, but is also in itself a representative of the message. Plotinos is clearly not writing of the Faustian 'summoning of demons from the vasty deep'; rather he describes a technique for concentrating or focusing the 'soul of the world' or Spirit of the universe, conceived as a third entity intermediate between the Body of the world (physical Nature) and the Intellect of the world, which is the divine mind. Ideas within this Intellect are reflected as 'seminal reasons' within the soul of the world, and these again engender physical bodies in the material universe.

Proclos, however, seems to have known of more curious practices involving the sacrifice of animals whereby the ancient priests

received, in the first place, the powers of daemons, as proximate to natural substances and operations; and by these natural substances they convoked into their presence those powers to which they approached. Afterwards, they proceeded from daemons to the powers and energies of the Gods ...

And lastly, laying aside natural substances and their operations, they received themselves into the communion and fellowship of the Gods. (Taylor 1818: 304).

A glorious vision indeed to come before the eyes of mystic philosophers in later ages.

The early Christians embraced neo-Platonism with enthusiasm. 'In the beginning was the Word, and the Word was with God, and the Word was God': the familiar opening of the Fourth Gospel (c. 100 AD) might come straight from a text of that school. The quiet morality of the Platonists, their modest and abstemious lives, their prudish attitude towards sex and their sense of the immanence of the Gods in the life of the mundane world struck sympathetic chords among the Christians. Several Fathers of the Church — Origen, Basil, Gregory of Nyssa — transmitted elements of neo-Platonism to the West, though sometimes amid severe criticism. Of them all, St Augustine (fourth century) was the most deeply immersed in the Platonic texts and in the earlier phases of neo-Platonic philosophy. He himself, it seems, first found the doctrine of the Word in those texts and their influence remained with him.

To Augustine, Plato's was the unique philosophy of the pagan world that was worthy of any admiration on the part of a Christian. 'No school has come closer to us [Christians] than the Platonists' he wrote, and again, 'we prefer the Platonists to all others' (St Augustine 1968, II: viii, 23; x, 49). His object, of course, was to show how far the Christian excelled over even the highest of pagan philosophies, most of which (especially those of the materialists and the atomists, with pre-Socratic origins) Augustine regarded as too contemptible to deserve refutation. At the end of his sketch of Greek philosophy Augustine declared: 'Plato is extolled because he ... perfected philosophy which he divided into three parts: *moral*, which is chiefly concerned with practice [i.e. conduct]; *natural*, which is devoted to theory; and *logical*, which distinguishes truth from falsehood.' He regarded this praise of Plato as just, while making it clear that for the Christian it is the Platonists' recognition of the true nature of the Deity (so far as such recognition was open to a pagan, unaided by revelation) that commends their philosophy to him. So, in natural philosophy, the Platonists rightly 'have said that the true God is the author of all things' and furthermore God's invisible attributes as well as his eternal powers and godhead were clearly seen and understood by them through created things'. In their logic, Augustine discovered that 'they have declared that the light which illuminates the intellects of men ... is this selfsame God', and in their ethics also the Platonists were to be admired because they reasoned that 'the true and supreme good is God' (1968, II: iv, 19; v, 27; vi, 35; vii, 37; viii, 43).

In whatever light we consider Augustine's willingness to set Platonism in the best possible view, to see Plato himself as the grand intellectual and moral hero of pagan antiquity and even to be willing to suppose that Plato had received illumination from Jewish sages whom he had encountered in Egypt, it remains true that for Augustine the Platonists failed because they were *not* Christians – of course a temporal impossibility for Plato himself and many of his followers. Christianity being so far in all respects above paganism, Plato's philosophy was simply the best of a bad lot. The neo-Platonists deserved stronger condemnation yet from Augustine because they had wilfully rejected Christianity and given themselves to the worship of 'demons'. Augustine devotes many pages (not his most limpid) to his exposure of the errors and absurdities of the 'Platonist Apuleius of Madoura', who lived in the second century AD and is best known to posterity for his picaresque fiction, *The Golden Ass*. Here Apuleius writes fervidly of the worship of the Egyptian goddess Isis. He also wrote a book on Plato, and another on *Socrates the God* (severely mauled by Augustine, who thought its title should have been *Socrates the Demon*), but Apuleius's chief offence to Christians was in his *Apology* for magic, a work which nevertheless seems to have succeeded in liberating its author from a charge of criminal practices of the magical kind. Apuleius's sin, in Augustine's eyes, was that he defended and praised demons, dwellers in the air between Gods (in Heaven) and humans (on Earth). Augustine no more doubted the real existence of demons than he doubted the existence of angels: his argument ran that demons had no access to the true God and that it was best for humanity to shun them and the magic they wrought: 'these magic practices without question do deadly harm to mankind.' Augustine held that demons could not be friends to humans because they were 'utterly vile and depraved', nor could demons be intercessors between God and humanity (1968, II: xix, 85; xxiv, 127). He argued at some length that the Platonists' conception of God was inconsistent with their toleration or cult of demons, whose chief delight was to revel in stage-plays depicting the Gods as ridiculous, in lewd ceremonies and in shameful cruelties – these being the worst aspects of pagan religious worship. 'What kind of folly, then, or rather madness is it to subject ourselves to the demons under the name of some religion, when true religion frees us from the vicious inclinations in which we resemble them?' (1968, II: xiv, 65; xvii, 81)

Another student of Platonism was Boethius (480–524), one of the crucial late-Latin authors with respect to intellectual life in medieval western Christendom (his *Consolation of Philosophy* was to be translated into English by King Alfred). In the history of learning Boethius is most often cited for his treatises on logic and the four mathematical arts (arithmetic, geometry, music and astronomy) which provided the bases for higher education in subsequent centuries. Boethius was a Christian, a martyr to his faith, but the philosophy of his last work (composed in prison) is not overtly Christian: it teaches the neo-Platonic doctrine that evil cannot prevail.

From Macrobius (a pagan, fl.395–423) the Middle Ages gathered the concept of perfection in the universe, another reflection of neo-Platonism. In this doctrine, there is only one truly perfect and immaterial Being, the One, the true God, found in the outer spaces of the cosmos. Moving inwards towards the Earth at the centre of the universe, materiality increases and perfection diminishes. The celestial spheres and the planets carried by them, beginning with Saturn (far less perfect than the encompassing sphere of fixed stars) become more and more material and degraded down to Mercury, taken to be the planet closest to Earth. The *Commentary on the Dream of Scipio*, in which Macrobius described this hierarchy of being, remained a familiar Latin text for schoolboy study from the Middle Ages to the twentieth century (Stahl 1952).

Also thought to have been influenced in some degree by the neo-Platonists was Boethius's North African analogue, the fifth-century compiler Martianus Capella. Less learned than Boethius, presumably not a Christian, Martianus's Latin digest of the mathematical sciences was, together with that of Boethius, the most important source of knowledge of these subjects in the West for many hundreds of years. The astronomical text has been characterized as the best in the Latin tongue to survive from antiquity: besides much that is very basic, Martianus explains Heraclitus' hypothesis that the planets Mercury and Venus revolve round the Sun rather than the Earth, about which the Sun and the three outer planets gyrate. This was virtually the only allusion to the Pythagorean tradition of astronomy known to the Middle Ages, and Martianus was duly praised for it by Copernicus.

Because Boethius was unable to realize his intention of translating all of Plato into Latin, knowledge of Plato's writings was slight and sketchy in the post-Roman world, indeed until the renaissance of the twelfth century opened a new age of translation. The *Timaeus* alone (partially translated in the fourth century AD) exercised a strong influence. This work, writes Klibansky in a masterly summary of the continuous line of medieval Platonism, 'with its attempted synthesis of the religious teleological justification of the world and the rational exposition of creation was, throughout the earlier Middle Ages, the starting point and guide for the first groping efforts towards a scientific cosmology' (1939: 28). To the *Timaeus* were added in 1156 Latin versions of the *Meno* and *Phaedo*. These three works, together with the *Commentary* on *Timaeus* by William of Conches (c. 1120) were well known to such humanists as Nicholas of Cusa and Nicholas Copernicus. Klibansky indicates a distinct anticipation of Henry More's later efforts in medieval 'attempts to harmonize the Platonic and Mosaic narratives and to interpret the Biblical account in Genesis by means of the Greek scientific categories and concepts which had become part of Western thought, mainly by way of the Latin Timaeus and its commentator' (1939: 28).

During the thirteenth century the doctrines of Christian scholasticism took shape, highly rationalist, predominantly Aristotelian rather than Platonic. Whereas, in the twelfth century, Platonism, neo-Platonism and Hermeticism had blended together, in the thirteenth the whole gnostic tradition of mystic philosophy and magic tended to vanish from sight, to reappear with the renaissance of art and learning in Italy. Nevertheless, it is Klibansky's firm opinion that love of the writings of Plato was continuously present, transmitted to the early Italian humanists, like Petrarch, to Cosimo de' Medici, and so to Marsilio Ficino, who was well aware of the medieval Platonists who were his precursors. '[T]he prevailing theory of a definite break between mediaeval and Renaissance Platonism which has dominated the history of philosophy cannot be maintained on closer inspection of the facts' (1939: 35).

Scholasticism was firmly opposed to magic and demonology, even stemming from so respectable a source as the neo-Platonist school. No less perilous were the arts of foretelling the future, as at least potentially diminishing the omnipotence of God and the free will of man. Even at the height of the popularity of philosophic magic (shortly to arrive) its masters were careful to insist that the star over Bethlehem and the eclipse of the Sun at the time of the crucifixion of Jesus were miraculous events beyond the skill of astrologers. Those who condemned magic or attempts to divine the future did not do so because they regarded the pretensions of magicians and astrologers as absurd, but rather because they were impious. Already in the second century Origen, whose works, like those of St Augustine, carry traces of his early training in the neo-Platonic philosophy, had castigated appeals to the demons made by sufferers seeking a medical cure, for such magical medicine, in effect, gave the demons power over our own bodies. Far better to pray for help to the true God, who can protect the righteous man against the demons; far better to use the formula in the name of Jesus' than to summon up demons.

For the same reason, St Augustine judged the doctrines and practices of the followers of the supposed ancient sage, Hermes Trismegistus (the 'thrice-great Hermes' of Milton), to be even more abominable than those of the neo-Platonists. Immediately after disposing of Apuleius and his kindred in *The City of God*, he turned upon their brother gnostics, no doubt fully cognizant of the connection between the Platonic and the Hermetic traditions traced by Lactantius (see below). Augustine particularly directed his arguments against the *Asclepius*, a work written by Greeks early in the Christian era in which Hermes is made to discourse at length. Augustine quotes copiously from it 'Hermes'' description of supposedly Egyptian temple magic, the magic to which (as we have seen) Proclos alluded later and which was held to offer a means of assimilating humanity to the gods. In Egypt (Augustine quotes the *Asclepius*), the priests make 'statues endowed with life, pregnant with sensation and inspiration, and performing many wonderful things; statues that have foreknowledge of the future, statues that can bring maladies on men and heal them again.' 'Hermes' explains later how this was done: 'since they [the priests] were unable to create souls, they called upon the souls of demons or angels and introduced them into the holy images' (1968, II: viii, 107; xxiv, 117). As he had argued against Apuleius the Platonist, Augustine again maintained that such arts were wrong and could do no good to men because of the malevolence of the demons; besides, he wrote, even non-Christians must be superior to those whom they have themselves fashioned as Gods. The created must be inferior to its creator: so what is the point of the Hermetic art? In any event, it is the fate of such statues to be overthrown by the victorious Christians and the demons cast out, as 'Hermes' himself had foreseen (with lamentations), according to Augustine's interpretation of the text.

Despite the opposition of the Fathers and other authorities in the Church to magical and neo-Platonic practices, one is by no means entitled to suppose that during the centuries between Bede and Dante learned people were free from astrology, alchemy, magic and superstition. Quite the contrary: belief in strange influences, powers and portents was universal. Everyone knew that the imagination of a pregnant woman could affect the infant she was carrying: a hare seen in a field could cause a harelip. Everyone knew of the salamander as the unique creature able to live in a fire; Benvenuto Cellini never forgot the great box on the ear given to him by his father on the day when they saw one, to help him remember the incident. Everyone knew that a comet foretold doom; that in Rhodes dogs fawned over Christians and barked at Turks; that frogs fell from the sky in showers; that garlic destroyed the attraction of a magnet; that geese were engendered from barnacles; that the ringing of church bells rendered thunder and lightning harmless. Everyone believed in the providential fitness of passages chosen at random from the Scriptures by one in a state of wretchedness, illness or perplexity. Popes and bishops blessed images for the protection of the faithful. Most people at least credited the virtue of pronouncing or writing special formulas – not only Christian prayers – especially if the original and primitive Hebrew language was used. Even those few who, like Pico della Mirandola, attempted publicly to discredit astrology seem to have been unable to rid themselves entirely of commitment to its assumptions. Arguments and instances in favour of the special involvement of the heavens in human affairs were almost universal in the fifteenth century. Marsilio Ficino noted that he and Pico had both been born under Saturn in the house of Aquarius, though thirty years apart; in the year of Pico's birth Marsilio had commenced his studies of Plato, by direction of Cosimo de' Medici; Pico himself arrived in Florence on the day of publication of Ficino's edition of Plato. The sympathy between them was no accident.

These two men were the leaders of neo-Platonic revival in Tuscany. The elder, Marsilio Ficino (1433–99), son of a physician at the court of Cosimo de' Medici, was introduced to the patronage of the Duke who supported him throughout his career as a translator and commentator on Plato and the neo-Platonists. From 1462 onwards he was head of the Platonic Academy assembled by Cosimo at Florence. Ficino was one of the second generation of Western Greek scholars. The council of Florence (1438–9), though it had failed to reunite the Greek and Roman Churches, had brought many scholars from Byzantium to Italy, notable among them Georgios Gemistos Plethon (1360–1452), who first aroused interest in the Platonic philosophy among the Italian humanists, and Cardinal Bessarion (1395–1472), whose attack on 'the calumniators of Plato' was to be printed in 1469. The search for ancient manuscripts was pursued with renewed enthusiasm, given fresh urgency by the fall of Constantinople in 1453. Seven years after this event a manuscript acquired in Macedonia was brought to Cosimo in Florence by one of the agents whom the Duke employed to seek out such rarities. Cosimo at once urged Ficino to set aside the translation of Plato that he was beginning in order to tackle this new manuscript, whose translation Ficino completed within a few months. It was a nearly complete text of the *Hermetic Corpus*, in all fifteen works attributed to Hermes Trismegistus. That done, Ficino went back to Plato – never had so many writings by the Master been put together before. The complete Latin Plato occupied Ficino for yet a further seven years; it was printed in 1484 and widely read by many generations. Scattered writings of Plato had been accessible before in Latin, but Plotinos was wholly new to the West. This translation also Ficino made between 1484 and 1486, and printed in 1492. Ficino also published commentaries of his own on both Plato and Plotinos, as well as books on many other subjects, theology among them. Until the (temporary) overthrow of the Medici family in 1494 he was also active as a teacher in Florence: indeed, the 'new Plato' was introduced to Western scholars by the lectures he gave at Florence, on the *Philebus*, probably in 1465. In the words of Paul Kristeller, 'the direct access to the work of Plato, of Plotinos and of the other Platonist philosophers of Greek antiquity was itself a major event in the intellectual history of Europe', despite the fact that the newly discovered writings of Plato himself were still confused with neo-Platonic and medieval ideas (Allen 1975: 2; Werkmeister 1963: 109).

The *Hermetic Corpus*, called by Ficino *Pimander* from the title of the first book in the collection, translated in 1463, was printed in 1471 and so for the first time Hermeticism received a wide currency in Europe. It had surfaced occasionally in the Middle Ages but for most scholars, until 1471, Hermes Trismegistus had been no more than a name to conjure with, whose writings were unavailable. The origins of the *Corpus* lie close to those of neo-Platonism, in the second century AD; indeed the mistaken tradition that coupled Hermeticism and Plato tightly together was confidently accepted by all the learned down to the seventeenth century.

The tradition had, indeed, been approved by respected Fathers of the Church. Lactantius – one of the few educated men known to have insisted on the flatness of the Earth – wrote of the Egyptian Hermes as a man of great antiquity, preceding in time both Plato and Pythagoras, who had written many books and understood something of the true God. Augustine, even more specifically, pointed to Hermes as the first moralist of Egypt 'long before the sages and moralists of Greece' though later than Moses. Ficino in turn slightly embroidered Augustine's words from *The City of God* (XVIII, xxix): 'in that time in which Moses was born flourished Atlas the astrologer, brother of Prometheus the physicist and maternal uncle of the elder Mercury whose nephew was Mercurius [Hermes] Trismegistus' (Yates 1964: 14). In more detail, using the same sources, Ficino, in his introduction to the *Pimander* translation, listed the succession of ancient sages (*prisci theologi*) as being Hermes, Orpheus, Aglaophemus, Pythagoras, Philolaus and 'our divine Plato'. Ficino repeated this intellectual geneaology many times, not always identically: sometimes Zoroaster (the legendary founder of fire-worship, whose modern exponents are the Parsees of India) also figures in it. Thus in the Hermetic historiography three schools of antique wisdom, those of Hermes, of Pythagoras and of Plato, were reduced to one, with Egypt, in the manner customary among the classical Greeks, given temporal priority as the fount of wisdom. Contrary to the teaching of Augustine, the Renaissance scholars chose to believe that Hermes had approximated more closely to Christians in his beliefs than had his successors. The texts which Cosimo de' Medici was so anxious to read in Ficino's translation therefore carried a certain religious cachet, as well as displaying (scholars then agreed) the oldest wisdom in the world, the true pre-Socratic wisdom of Egypt when it had been the resort of Thales, Pythagoras and Plato in their search for truth. (The Renaissance knew the names of some pre-Socratic philosophers, such as Democritos and Anaximander, besides Thales (all of whom were to be found in Aristotle), but paid little attention to them; or rather, only scholars captivated by Greek atomist ideas heeded them.)

Ficino already knew the Latin version of the *Asclepius*, attributing its authorship to Hermes. He found important similarities between this text, the teachings of the neo-Platonists which he had rediscovered, and the works of the *Hermetic Corpus*. He therefore accepted the categoric statements of Lactantius and Augustine that Hermes Trismegistus had furnished inspiration to Plato and his followers. We now know – since the investigation of Isaac Casaubon in 1614 – that this temporal order is the reverse of the true one, but it was highly plausible in itself, fitted many of the prejudices of the ancients (for example, regarding the Egyptian influence upon Greek philosophy and medicine), and consorted with the history of Christianity. Down to the seventeenth century Ficino's interpretation, his view of the *Hermetic Corpus* as 'the pristine fount of illumination flowing from the divine *Mens*, which would lead him to the original core of Platonism as a gnosis derived from Egyptian wisdom' was universally endorsed (Yates 1964: 17). And as Frances Yates indicated, the chronological misapprehension entailed an interpretative misapprehension also: for Renaissance scholars could not possibly conceive of Hermeticism as a late and possible degenerate derivative from neo-Platonic philosophy. Rather they were confident that in the older wisdom of Hermes was to be found the essential key for understanding the subsequent philosophy of Plato. And it must be added that long after Casaubon had restored the correct chronology and sequence of the texts, the authority and prestige of Hermes remained unshaken. Certainly Henry More treated the Hermetic writings with profound respect.

As Frances Yates has explained, Ficino's discovery of Hermeticism as the seedbed of Platonism set magic in a new light. This may well have suited the spirit of the times. Perhaps affected by the disastrous events of the fourteenth century – above all the demographic collapse associated with the Black Death from 1345 onwards – the fifteenth century showed a growth of credulity towards mysteries and the occult, including witchcraft, magic and alchemy. One of the most famous or infamous of witchcraft trials, that of Joan of Arc, took place in 1431. This growth of credulity the full flowering of Renaissance scholarly humanism and the study of Greek and Latin scientific texts that flowed from it. It penetrated the universities, where the dependence of medicine upon and the subservience of astronomy to astrology seem both to have increased. Ficino's *Book Concerning Life* (1489) was a frequently reprinted manual about the preservation of health and the treatment of disease on the basis of the relationship between the parts of the body and the signs of zodiac and the planets

(1964:62). Ficino was also a strong advocate of the use of talismans. Circumstances, including the judgement of the Church, were therefore more favourable for a revision of that universal condemnation of the occult and magic arts which had been carried down from the time of the Fathers, and sometimes enforced. What is in question here is not of course popular occultism – witchcraft or any form of demonism – which was always cruelly suppressed (when discovered) but a scholarly or Faustian magic which was far more often expressed in speculations than realized in actual operations. The improved status of such scholarly or Platonic magic

> was mainly due to that great flood of literature which came in from Byzantium, so much of which dated from those early centuries after Christ in which the reigning philosophies were tinged with occultism. The learned and assiduous reader of such authors as Iamblichos, Porphyry or even of Plotinos, could no longer regard magic as the trade of ignorant and inferior persons. (1964: 17–18)

Thus magic rose up the social scale. It also, inevitably, tinged mathematics with a deeper dye of mystery from the series of equivalences: mathematics equals Platonism, Platonism equals Hermeticism, Hermeticism equals magic. To mystically minded scholars of the Renaissance this was an advantage in that they could occupy themselves harmlessly with complex diagrams, rich in symbolic meaning (such as those devised by the English Hermeticist Robert Fludd (1574–1637)) or with abstract symbols supposedly endowed with great power (such as the hieroglyphic monad of the English esoteric mathematician, John Dee (1527–1608)). The majority of the scholars who took this course contributed little either to the recovery of Greek mathematics – the springboard for Europe's mathematical progess – or to the actual development of arithmetic, geometry or algebra. An exception was the Italian mathematician and philosopher Girolamo Cardano (1501–76) whose fame is as great in the history of algebra as in the history of occultism. Equally, the majority of the creative figures in mathematics proper during the sixteenth and early seventeenth centuries were neither Platonists nor magicians. In the popular mind of the time, however, the association between mathematics and magic was inviolable; geometry became a 'black art'. The association was of negative value to real mathematics and sometimes brought distress to those whose beliefs and activities furthered it, as when Dee's library was sacked by a London mob in 1583.

Scholarly defenders of magic against ecclesiastic and popular prejudices employed a variety of argument. One, adumbrated already, tried to cover the defects of Hermeticism by the acknowledged merits of Platonism, its supposedly lineal descendant. How could anyone imagine the divine Plato to have been stained by a vile resort to demons? St Augustine had given to Plato the highest praise a pagan philosopher could merit. St Dionysius the Areopagite, the saint whom St Paul had encountered at Athens, was, according to St Thomas Aquinas as well as Marsilio Ficino, the author of *Celestial Hierarchies*, a Christian but highly neo-Platonic work, harmonizing the notions of the Platonists with a scheme of nine orders of angels in three ranks. (Here was another learned error: the work was not by the Areopagite but by an unknown author some centuries later.) Its argument fails to meet the direct criticism of the Platonists by Augustine.

Secondly, Ficino tried to demonstrate the similarity between the expressions of Hermes and those of Christians, and so of the underlying thought. He had before him the example of Lactantius, who had long ago noted in the Greek *Asclepius* that the Hermetics spoke of God as the Lord and Creator of all things, who had also made a second God, whom he hallowed 'and altogether loved Him as His own Son'. Ficino drew attention also to the close resemblance of the Hermetic account of the creation of the world to that given in *Genesis*: he refers to the cosmogony of Plato in the *Timaeus*, then goes on: 'Trismegistus Mercurius [Hermes] teaches more clearly such an origin of the generation of the world. Nor need we wonder that this man knew so much, if this Mercurius was the same man as Moses, as Artapanus the historian shows with many conjectures' (1964: 7, 26). Ficino became more than half persuaded of the truth of this identification. He also examined the analogy between an 'Egyptian' description of the Fall of Man and that lying at the root of the Hebrew-Christian religion; he found that the 'Egyptians' had distinguished an earthly part in humanity from a part that was divine (in truth, demonic). In the Hermeticists he discovered, in addition, an inkling of the regeneration of humanity, and he regarded the *Asclepius* as a 'divine' work on the will of God, even though this text contains the clearest references to the infusion of demonic power into statues to make them come alive. Ficino endeavoured to Christianize Hermeticism by turning a blind eye to its dangerous aspects, thus by implication liberating neo-Platonism also from the evil taint of demonic magic. In *De vita coelitus comparanda* he put forward 'a programme for non-demonic magic, utilizing the *spiritus mundus* and reaching no higher than the human spirit'. But elsewhere he addressed magic 'to good planetary demons' and it seems certain that he tested the efficacy of talismans, despite the warning of St Thomas Aquinas against such commerce with (possibly malevolent) intelligent beings (Walker 1958: 43; 53). Recent scholars reject the suggestion of Arnaldo della Torre (1902) that the conflict between his religion and his Platonic philosophy drove Ficino into a nervous breakdown during the 1460s. There are really no grounds for supposing that such departures from orthodoxy as he made disturbed his peace of mind. He regarded the restoration to Western Europe of knowledge of the ancient wisdom of Plato and Hermes as a task ordained upon him by God.

Thirdly, Renaissance scholars emphasized the distinction between black and demonic magic, which they would condemn as hotly as anyone, and 'natural' magic, to be treated as legitimate. There were two levels of natural magic. At one level the practitioner exploited ordinary, but unfamiliar, properties of Nature in skilful ways in order to bamboozle the uninitiated: this was no more magic in the true sense of the word than are the adroit illusions of a stage magician. Elaborate physico-mechanical devices to arouse wonder in the minds of the crowd had been fully described by Hero of Alexandria (first century AD), such as the opening of the temple doors without the aid of human hands when the fire is lit upon the altar. The tradition of wondrous automata continued in Byzantium and the Islamic world and fragments of it passed into Western Europe, perhaps influencing the invention of the mechanical clock. The quality of the results achieved by practitioners depended upon their knowledge of the real properties of things, and their manual dexterity; optical illusions also were within their power and the mysterious changes wrought by chemical reactions, and (in time) magnetism could be exploited also. Such natural magic largely fills the book of that name by Giovanni Baptista Porta of 1558; a late relic of similar sort is the *Mathematical Magic* by John Wilkins (1648). At the other level, the natural magicians claimed to exploit truly secret and hidden powers of nature ordinarily employed by none but nevertheless contained in the normal order of things. Like the former, there was (they said) nothing demonic or improper in these natural powers, the only mystery being in the practitioners' knowledge of their existence and use. Electrical attraction may serve as an example of such an 'occult' force that does really exist; however, the practitioners also invoked, with the greatest confidence, powers of which we are now sceptical, powers involving strange sympathies and antipathies and action at a distance. Some call to mind a widespread superstition that the recent wounds of a murdered person will bleed afresh if the murderer touches the corpse. Such an effect was not thought to be demonic: it was simply a consequence of the revulsion of the moral order of Nature against a dreadful crime, like the shriek of a mandrake when it is pulled from the ground. A famous medical instance of the mid-seventeenth century was the 'powder of sympathy' popularized by the eccentric philosopher, Sir Kenelm Digby (1603–65). The powder, applied to the weapon that had caused a wound, effected a cure of that wound, however remote it might be. The cure was thought to result from the proper use of the occult but genuine sympathy between natural things which was as real to Digby as the 'sympathetic' vibration between two strings in tune, or as universal gravitation was to Isaac Newton.

Natural magic touched upon the doctrines of orthodox philosophers in a variety of ways. In the first place they had to admit that there were latent in Nature phenomena whose cause was still unknown, 'occult' in the simple sense, hidden because undiscovered. An unusually skilled person, therefore, exploiting recondite knowledge, might work natural results or bring about medical cures that would appear to be products of sorcery only to the ignorant. So the Aristotelian philospher Pietro Pomponazzi wrote in 1556:

> if it is true ... that there are herbs, stones, or other means of this sort which drive away hail, rain or winds, and that one is able to find others which have naturally the power of attracting them ... it follows that men are able, by applying the active to the passive, to induce hail or rain and to drive them away: for myself, I see no impossibility in this. (Ron Millen in Osler and Farben 1985: 193)

Pomponazzi's confidence has not so far been redeemed by experience; had he used the communication of information from one place to another far distant – another application of natural magic often recited in the sixteenth century – his prediction would have been verified.

Secondly, in the spirit of Plato against that of Aristotle, natural magic strengthened the idea of the existence in Nature of structures, properties or powers that are inaccessible to ordinary investigation. The traditional

Aristotelian expected reality to be consistent at all levels: however finely cheese, chalk or star were divided, cheese, chalk and star they would never cease to be, always distinct one from the other. There was no point at which the Aristotelian philosopher must admit that ours is a looking-glass world, mirroring reality. In holding that the world of experience is no more than a reflection or image of a real world of existent entities that escape our senses, the Platonist joined hands with the atomist – or what the seventeenth century called the mechanical philosopher – though of course there was no similarity at all between the Platonists' conception of reality and the atomists'.

Thirdly – and this point too is illustrated by Pomponazzi's words – natural magic persuaded some philosophers (including Francis Bacon, according to recent scholars) to take a more purposeful view of the investigation of Nature: the aim might be not only to know, but to control. The object of all magic, demonic as well as natural, was to give the magus greater power over the world, with consequential prestige and authority, whether to command the weather (as Prospero in *The Tempest* is supposed to do), or to reveal precious mines of minerals unsuspected below the soil. Magic was an active art rather than a contemplative science. Though the great French historian of science, Alexandre Koyré, and others of his generation, thought that this emphasis on the mastery of Nature was an aberration, irrelevant to the true cause and progress of the Scientific Revolution, other scholars since have taken the opposite point of view, arguing (for example) from the frequent claims for the present and future usefulness of scientific knowledge made by members of the scientific academies that were organized in many countries late in the seventeenth century. If the utilitarianism of magic is judged to be a factor influencing the evolution of 'rational' science, its origins were certainly Hermetic rather than Platonic in the strict sense. Plato's ambition was to give people command of themselves, not of the natural environment. However, in so far as his successors accepted the idea of priestly power and temple magic, they gave countenance to the more forceful notions of the school of Hermes Trismegistus.

Finally, the idea of 'sympathy', whose manifestation we may crudely call 'action at a distance', is one with a long and curious history. Apart from psychological phenomena – two minds thinking or feeling alike, the sense of like or dislike between individuals – there were many instances to be quoted from the physical world. If one of two identical strings is plucked, the strings not being very far apart, the second will vibrate 'in sympathy'; any phenomena that we associate with resonance would fall into this class. Objects placed near the fire grow hot without contact with the flame; a magnet attracts one end of a compass needle, repels the other. Natural history too offered many instances dear to the encyclopaedists, of sympathy and its opposite, antipathy; the fatal attraction of the candle for the moth, the antipathy of man for snake, of cat for dog. The conviction that like attracted like, and unlike things repelled each other, was so firmly rooted that the philosopher Cardano stated the rules of magnetic attraction precisely in this sense: like poles attract, unlike poles repel, which is false. Some philosophers believed that attractive and repulsive effects, apparently occult, could be explained by appeal to the inaccessible micro-world: thus, if fire ejected heat-particles violently, these would enter remote articles and warm them. Emitted, invisible and indetectable effluvia might explain magnetic and electrical effects mechanically: so, it was pointed out, it was necessary to rub amber (in the one case) or stroke an iron bar with a lodestone (in the other) to set the effluvia working. The great Newton chose to consider magnetism and electricity as effluvial, or the work of subtle fluids, distinguishing these attractions from that of gravity. The physician Girolamo Fracastoro had in 1546, somewhat in the same vein, attributed sympathy and antipathy to the propagation of 'species' (undetectable forms or simulacra of objects) into the neighbourhood of the active bodies; this was simply an extension of a medieval theory about the radiation and reflection of light and the eye's vision of objects. Others still thought that sympathies and antipathies were not to be explained away but to be regarded as genuinely occult effects in Nature.

Long before the time of Bacon and Galileo the endeavour of Ficino to free hermetic philosophy from the taint of paganism had fully succeeded. His writings were to open 'a floodgate through which an astonishing revival of magic poured all over Europe' (Yates 1964: 61). Already, before the end of the fifteenth century, the sage Hermes Trismegistus had become so venerable, so benevolent, so Christian that in the mosaic pavement of the *Duomo* of Siena he is portrayed as the herald of Christianity, almost as though he had been one of the old Hebrew prophets or John the Baptist himself. Ficino's belief in the efficacy of appropriate images, painted in the correct colours and worn as talismans or else used to adorn the ceiling of one's bed-chamber, in order to secure good health and long life, no longer seemed superstitious or counter-religious. Nor did it now seem grievously improper to sing the ancient songs of Orpheus to the lute's accompaniment, or to try the magic power of Hebrew words, learned from the Cabala.

Of the effigy at Hermes of Siena Frances Yates has written:

> The representation of Hermes Trismegistus in this Christian edifice, so prominently displayed near the entrance and giving him so lofty a spiritual position, is not an isolated local phenomenon but a symbol of how the Italian Renaissance regarded him, and a prophecy of what was to be his extraordinary career throughout Europe in the sixteenth century and well on into the seventeenth century. (1964: 42–3)

Though not all philosophers were Platonists and not all Platonists magicians, the derivation of the magical Hermetic tradition from Platonic philosophy was firmly and repeatedly stated by its Renaissance adherents. Thus Henry Cornelius Agrippa (1487–1535) listed Pythagoras, Plato and all the most eminent Greek neo-Platonists in an enumeration of the greatest magicians, along with Hermes, Zoroaster and Orpheus (1964: 131). It was to be a distinction of the Cambridge school of Platonists that they departed from this magical tradition.

The split between the Cambridge Platonists and Platonists of the Hermetic and magical tradition is well illustrated by the literary quarrel of 1650 to 1651 between Henry More and the alchemist Thomas Vaughan (1621–66) (Burnham 1974; Rudrum 1984). Although an extravagant writer, a Rosicrucian and mystic, Vaughan seems to have possessed genuine chemical skills which made him a valued friend of such founder Fellows of the Royal Society as Sir Robert Moray and Thomas Henshaw. If the report that he died of mercury poisoning be true, he was a martyr to alchemy. Because its practitioners were in general such unworthy men as Vaughan, chemistry was the experimental science least esteemed by More, while Vaughan, in true Hermetic vein, claimed to be a better Platonist than he: 'thy *Ignorance* and Insufficiency in the *Platonick* Philosophy' was manifest in More's philosophical poems (1647). 'Thou didst fancy thy *Psychodia* for a rare profound piece, and that *Timaeus* was inferior to thy *Coplas* [couplets].'

Hence Vaughan's urge to correct the slighting of the true, esoteric Platonic tradition by this

> *Master* of *Arts* of *Cambridge*, and ... *Poet* in the *Loll* and *Trot* of *Spencer*. It is suppos'd he is in *Love* with his *Fairie-Queen*, and this hath made him a very *Elf* in *Philosophie*. He is indeed a scurvie, slabbie, snottysnouted thing. Hee is troubled with a certain *Splenetic loosnes*, and hath such *squirts* of the *Mouth*, his Reader cannot distinguish his Breath from his *Breech* ... But I have studied a *Cure* answerable to his Disease, I have been somewhat *Corrosive*, and in defiance to the *old Phrase*, I have washed a *Moore* clean. I have put his *Hog-noddle* in *pickle*, and here I present him to the World, a *Dish* of *Sous'd Non-sence*. (Rudrum 1984: 242).

Such punning vituperation is typical of Vaughan's invective. More, who began the quarrel, was no loftier in his language (if not altogether a match for his adversary), calling Vaughan a Fool in a Play, a Jack-pudding, a giddy phantastic Conjuror, a poor Kitling, a Philosophic Hog and a Meg with a Lanthorn (among other things). The argument about Platonic philosophy is hard to discern beneath the name-calling; one can just discern that More thought Vaughan a dangerous fraud in his talk of Agrippan and Lullian mysteries: 'I have always [Vaughan had written] honoured the *Magicians*, their Philosophie being both *rational*, and *Majestic*, dwelling not upon *Notions*, but *Effects*, and those such as confirme both the *Wisdome* and the *Power* of the *Creator* [and so I] found at last that *Nature* was *Magicall*, not *Peripateticall*' (1984: 215).

While equally convinced that Nature is not peripatetical, and also believing in a 'Spirit of Nature' (though not meaning the same thing as Vaughan) More offered no concessions to magic. On his side Vaughan, rebutting More's accusations of irreligion, contended that More cavilled about matters beyond his understanding. He would admit that the 'broyling frying Company, who call themselves Chimists ... are indeed no true Philosophers'; nevertheless Vaughan was confident of the powers of the philosophical fire and the virtues of the philosopher's stone. He continued to assert 'the *truth* of that *Science*, which I have prosecuted a long time with *frequent* and *serious indeavours*', a science based on the cabalistic learning found in Dionysius and Zoroaster, Patrizi and Basil Valentine, Pistorius and many more (1984: 313; 340; 359; bibliography).

That the Cambridge Platonists rejected the Hermetic tradition, for all its supposed derivation from the philosophy of the Platonic school, did not entirely result from Casaubon's exposure, early in the seventeenth century, of the falsity of the claims for the hoary antiquity of Hermes Trismegistus, for Casaubon's rather unemphatic falsification was accepted only slowly. More never mentions it. Of greater importance was religious doubt. In the unpublished section of his biography of More, Richard Ward wrote, of his attacks on Vaughan, that 'The design of these was to discountenance, as he tells us, *Vanity* and *Conceitedness*; and that *phantasy* and *Enthusiasme* wch had got by this time, in those days of Confusion, into Philosophy, it should seem, as well as Religion: and of wch this Gentleman was an egregious Instance' (Christ's College, Cambridge, MS). Also significant was the reception among the younger Cambridge men of the mathematical and mechanical philosophy promulgated by Galileo and Kepler, Descartes and Marin Mersenne. Mersenne (1588–1648), friend and coadjutor of Descartes, was an effective Christian apologist who firmly, indeed scornfully, refuted occultism and Hermeticism as represented by the English physician Robert Fludd (1574–1637) and the Italian philosopher Tommaso Campanella (1568–1639) (Lenoble 1943). No less anti-occult was the 'Pythagorean' astronomer Johannes Kepler. Astrology was slowly ceasing to be respectable – though both Kepler and Galileo drew up horoscopes on occasion – as astronomy and medicine drew away from its dubious support. Occult causes were officially banned from mathematical and experimental philosophy, or as some would say were transferred to the unattainable world of microphenomena, where nothing was verifiable or falsifiable. Nevertheless, occult and Hermetic ideas and practices survived in modified forms, though less and less important features of the main currents of philosophy and science. Just as Platonism (at least in Cambridge) separated itself from Hermeticism, so Hermeticism seemed to become more independent and idiosyncratic in its course.

Alchemy, for example, had played little or no part in the Renaissance philosophy of Marsilio Ficino and his successors; it reached its high point in the seventeenth century, side by side with the scientific revolution, when (besides many enthusiastic authors from Alexander Seton and Michael Sendivogius through J. R. Glauber, Thomas Vaughan and William Starkey) natural philosophers with wider perspectives and profounder purposes such as Joan Baptista van Helmont, Robert Boyle and Isaac Newton, each in his individual way, believed that the experiments and understanding of the alchemists might offer important contributions to the systematic study of Nature. It was hardly before the eighteenth century that writers on chemistry began generally to denounce the Hermetic legends of the origin of their science among the priests of Egypt and put aside all hope of effecting the transmutation of metals or winning the philosopher's stone. Overt alchemists were, of course, excluded from the rising 'establishment' of experimental science everywhere while those within it who were fascinated by its potentialities (with varying degrees of scepticism) were careful to keep their interest from the public view. A crucial point here (if perhaps one that is strictly irrelevant to the Cambridge Platonists) is whether students of the alchemical writings believed the knowledge and processes supposedly contained in them to be sufficient in themselves, because when understood they would put the philosophers' stone or transmutation within their power, or whether they hoped rather to gain from their study not such power, but the deep understanding of the workings of Nature that the power implied. Boyle and Newton belonged (I believe) to the latter group. In the end nothing resulted of benefit to science from the esoterica in the alchemical writings, despite all the effort to find a rational meaning in them: as Newton wrote to Locke in 1692: 'I have forborne to say anything against multiplication in general because you seemed persuaded of it, though there is one argument against it which I could never find an answer to' (Newton 1959–77, III: 218–19). In the Age of Enlightenment both astrology and alchemy became mere superstitions.

Hermeticism lingered in other ways too, in the esoteric Brotherhood of the Rosy Cross (if this was ever more than a fiction), in the first steps of Freemasonry, and in other semi-secret, semi-religious groups; in the chemical philosophy of J. B. van Helmont and still more in the ideas of his mysterious and strangely magnetic son Francis Mercurius (so aptly named!), friend of Henry More and of Leibniz, both of whom respected him immensely. Hermeticism lingered also in the reluctance of many scholars to dispense, after Casaubon, with the concept of the ancient sages (prisci theologi) as pioneers of philosophy and founders of science. The idea of a Golden Age at the beginning of the world, of which this was an element, was abandoned with regret.

PLATE 1 Double portrait of J. B. and F. M. Van Helmont *from Joan B. van Helmont's 'Ortus Medicinae' 1648. (Reproduced by kind permission of The Bodleian Library, Oxford: ref: Lister C.90)*

3

Platonism and the Scientific Revolution

THE SIMPLEST and least troublesome definition of a Renaissance Platonist is 'a follower of Marsilio Ficino'. Probably everyone to whom the term might be applied studied Ficino's books and his translations of Plato's writings. Otherwise there were wide variations within the group. P. O. Kristeller, doyen of Renaissance scholars, offers a conception of a Platonism that was placid and banal, engaged in 'establishing a kind of rational metaphysics beside, rather than against, dogmatic theology and empirical science. The Platonist conceptions that were most popular in the sixteenth century were the doctrines of the contemplative life, of immortality, of the dignity of man, and of Platonic love' (1979: 162). Others, with whatever justice regarding themselves as Platonists, leaned affectionately towards esoteric and magical arts. Besides the Platonist followers of Hermes Trismegistus, who might or might not be active practitioners of alchemy, there were the hardly less numerous students of the magical mathematical arts, who equally took themselves to be in Plato's tradition. The names of John Dee (1527–1608), Giordano Bruno (1548–1600) and Robert Fludd (1574–1637) are well known among the circle of these active near the turn of the century. Equally far from numerology and cabalism, talismans and alembics, were those for whom Plato had provided that essential key to the study of natural philosophy: knowledge of geometry. They believed that the architecture of Nature is rationally, not magically, mathematical. Late classical authors, rediscovered during the Renaissance, provided a Platonic antithesis to the mysticism of the neo-Platonists. *On Architecture* by the Augustan engineer and technical expert Vitruvius is an outstanding example of this genre. Vitruvius expounded to the Renaissance and later centuries the philosophy of mathematical proportionality as the foundation of aesthetic satisfaction (in pictures and buildings), and equally of technical perfection in all those works wherein man supplements Nature (from waterworks to missile engines). The development of this philosophy, at once artistic and practical, so evident in Leonardo da Vinci, went far beyond any original Platonic hints. To it we may also relate the 'Platonism' (if the term is indeed appropriate) of Copernicus, Kepler and Galileo. But their mathematical studies had also a different inspiration: the technical writings of antiquity on astronomy, optics and mechanics.

We have always to remember that the immediate intellectual context even of this most rational-seeming aspect of Platonism, where philosophy borders on technical astronomy, admitted ideas shared with the esoteric aspects. So Copernicus (in Book I of *De Revolutionibus*) applauds Hermes Trismegistus' calling the Sun a visible God; Kepler believed to the end of his days that the invisible God had fashioned the (finite) universe after the model of the five regular solids (or Platonic bodies). Among ordinary people the beliefs that the heavens are the seed-bed of forms, and that from them originate hidden virtues working secretly upon all things here below, were as commonplace in the sixteenth century as the conviction that disembodied human souls can animate gross matter. Some historians have maintained that credulity and superstition increased during the Renaissance: 'the turning away from Aristotle's philosophy and the rise of Paracelsanism encouraged the development of occult philosophy and a favouring attitude towards natural magic' in the opinion of Lynn Thorndike (1941, V: 14). On the other side, it might be argued that the Aristotelian philosophy was far from being a dead subject – we have noted its revival in Padua – and Paracelsanism for all its defects was more than a new occultism.

Some Platonist scholars of the sixteenth century understood clearly that the authentic Plato had no more advocated magical or cabalistic practices, of which indeed he knew nothing, than had Aristotle. Some, such as Francesco Patrizi (1529–97), who held specially created professorships of Platonic philosophy at Ferrara and Rome, were notable for their contributions to mathematics and philosophy *sensu strictu* (see below). Platonism in the broadest sense was not even restricted to moral and natural philosophy: the importance for political thought of Plato's *Republic* was well known and his ideas of literary form too were influential:

> As regards poetry and imaginative literature in general, the critics of the Renaissance appealed from the Plato of the *Republic* and the *Laws* to the Plato of the *Ion,* the *Phaedrus,* and the *Symposium.* Beauty being the subject-matter of art, Plato's praise of beauty was transferred by the Renaissance to poetry, and his praise of the philosopher was transferred to the poet. (Spingarn 1920: 156)

Taking account of the breadth and variety of Platonic influences, it is perhaps not surprising that attempts to summarize briefly only those elements in Renaissance Platonism that relate to the phenomena of Nature should strike the reader as tending to emphasize either the bizarre or the commonplace. It may seem easier to state the beliefs that Platonist philosophers did *not* hold, rather than the positive tenets of their philosophy. They did not believe (as many in the Renaissance supposed Aristotle to have done) in a unique, indivisible, immortal world-soul; likewise, they rejected materialism in every form, and every attribution of the nature of things to the operations of chance. They did not accept attempts to explain away wonders and marvels apparently contravening the ordinary course of Nature on the grounds that stories of such things were all false or else to be accounted for by the action of such known powers as magnetism. They did not believe that the categories of academic philosophy were adequate to embrace the world as we experience it. They denied that the natural world was so far beyond human control or influence as academic philosophy claimed. And perhaps the most important negative point of all: Platonists denied the assertion (variously attributed to Avicenna or Averroes) that there is no spiritual existence – in addition to the Creator – besides human souls and celestial intelligences. Platonists, following the principle of plenitude, were on the contrary convinced of the existence of a whole population of spiritual beings intermediate between humanity and

God: not only the several ranks of angels and archangels, but many other kinds of spirits and demons also.

In this last respect Platonism flourished by exploiting the hiatus between Aristotelianism and Christianity. The long-continued endeavour to Christianize the thought of the great pagan philosopher had never succeeded in giving total conviction to every honest reader. It was as obvious that Aristotle had rejected divine creation of the universe as that he had been ignorant of the saints and their miracles. Platonic (or neo-Platonic) spiritualism was far more easily reconcilable with Christian dogmas; the inherent (but at first largely disregarded) danger was that the process of such a reconciliation might entail the authentication of every tale of magic, ghosts and witchcraft. It is therefore not surprising that Aristotelians, such as Pomponazzi, were, on the whole, much less credulous than were the Platonists.

When, in the early seventeenth century, the 'new philosophy' began to replace in the allegiance of the younger and more daring scholars the 'old philosophy' of Aristotle, the situation was not fundamentally altered, since the 'new philosophy' was no less productive of prickly points of difficulty with respect to theology than Aristotelianism had been – indeed, as is well known, the philosophies of Galileo and Descartes were widely held to be far more damaging to religion than the now comfortably adjusted doctrines of the Schools. Galileo's advocacy of the Copernican ordering of the heavenly spheres and of the motion of the Earth as a planet occasioned the worst collision between faith and reason since the Archbishop of Paris had condemned Aristotelian philosophy in 1277. The Galileo crisis was, however, largely a problem for Roman Catholics. Protestants of all sects, little as Copernicus had commended himself to Luther and Calvin, were, by the mid-seventeenth century, able to congratulate themselves (rather smugly) on their more progressive intellectual attitudes and willingness to tolerate scientific radicalism. But almost all Christians, philosophers or theologians, were distressed by the revival of materialism, whether in the form of Epicureanism – that is, revived Greek atomism – or of the distinct 'mechanical philosophy' of Descartes. The degenerate epithet 'epicure' retains to this day a pejorative taste and Epicureanism as a philosophy has rarely enjoyed a good press: it has too often been judged to lead to atheism, a condemnation repeated over and over again in the seventeenth century. Epicureanism entailed the exclusion of God and all lesser spiritual beings from the universe, robbing man of his spark of soul and hope of salvation. Therefore it was a principal task of all those who favoured the mechanical philosophy, and particularly atomism, to free that philosophy from this dreadful taint by proclaiming that atomism was no more than a theory about the nature of matter pretending to no refutation of God's creation of the atoms and of the laws of motion governing their various combinations. Thus in a stanza from a poem by Henry More himself:

> And to speak out: though I detest the sect
> Of *Epicurus* for their manners vile,
> Yet what is true I may not well reject.
> Truth's incorruptible, ne can the style
> Of vitious pen her sacred worth defile.
>
> ('Democritus Platonissans', in *Philosophical Poems* (1647),
> quoted in Gabbey 1982: 182)

It may seem paradoxical that the chief restorer of atomism in the seventeenth century, Pierre Gassendi (1592–1655), was a Catholic priest and that Robert Boyle and Isaac Newton, so prominently associated with the development of atomism in England, were also men of outstanding piety. Descartes, too, was sincerely concerned to maintain his own good standing in his Church, though he inhabited the Protestant Netherlands by preference. A million contemporary words must have been written in denunciation of every avowed atheistical materialist in that age, so tiny was their number. But it was and is undeniable that natural philosophy became more materialistic in the seventeenth century than it had been under the domination of Aristotle, in whose philosophy the self-regulating, self-perpetuating, internally harmonious properties of the universe rendered it a gigantic organism. As Galen wrote, things were as they were because it was their nature to be such, and it was in the nature of things that the perfection of the part should bring about the perfection of the whole.

The seventeenth-century modernists adopted a wholly different metaphysics. Thrusting aside teleology and the maintenance of perfection, deliberately shelving the problem of the origins of the universe (or rather adopting the solution that rang most in unison with the Book of Genesis), they accepted the view that the universe we observe is *de facto* constant and stable: whether or not the universe was tending towards some gradual impairment of its order, or liable to destruction by the divine fiat, were rather theological than philosophical problems. Order, constancy and regularity in phenomena, which of course comprised much cyclical change such as the coming-into-being and passing-away of living things, were maintained (always under the divine providence) by the internal harmony of the structure of the universe, just as the regular going of a clock is secured by the proper proportioning and ordering of its wheels. By the middle of the seventeenth century, virtually all thinking philosophers – not necessarily the pedagogues and apologists for the past – were confident that the macrostructure of things (the 'secondary qualities' by which we distinguish sugar from salt, sheep from tree) is illusory, and that the permanent reality of Nature is its microstructure ('primary qualities' of magnitude, mass and speed) which considered on the cosmic scale remains constant, though subject to local variation and often reversible (as when water boils off as steam, or soil, air and water go to make a plant). Philosophers realized – in time – that the proportion of 'solid matter' (particles) is minute in proportion to the volume of 'space' between the particles. Following such general principles of the 'mechanical philosophy', it made little difference whether the philosopher believed or denied that there are ultimate indivisible particles (atoms), or that the particles somehow act directly upon each other across an empty, or inert space (vacuum) or are acted upon by an agent (God, aether, field) filling all space. The decisive philosophic issue was the conception of a systematic structure of matter in which all change (chemical transformations, the transmission of heat, the revolution of a planet about the Sun) is brought about by processes wholly internal to the system (as one can grow a little garden enclosed in a sunlit glass jar). Mysterious powers, spirits, intelligences, vital forces, demons – all such entities were banished. Descartes made a point of the greatest philosophic and scientific importance when he deduced from the essential constancy of the universe the constancy of the amount of motion within it (we have improved on Descartes by substituting the concept of energy for the more restricted concept of motion). For Descartes, things on average are neither slowing down nor speeding up, and no process can either add to, or detract from, the total amount of activity in the universe.

This philosophy was called 'mechanical' because its ultimate entity is the particle, or unit of matter, and its ultimate activity is motion, that is, the displacement of particles with respect to one another and the partition of activity by the impact of particle upon particle. To avoid the charge of materialist atheism, seventeenth-century philosophers not only invoked God as the eternal, all-creative Mechanic or Clockmaker but allowed human beings a supernatural status: only human beings possess an immortal soul and free will. Though their body might be constrained by the laws of mechanism, their intelligence is not. Descartes insisted that the body is a machine, in humans uniquely inhabited by a mind/soul capable of exercising free will against the appetites that dominate the animal. From this doctrine it was clear that spirits as such had no place in the universe; the long-receding frontier of animism had been pushed right away. The spirits of grove and stream had gone first, then the planetary intelligences, now at last humanity alone retained possession of a spiritual nature.

This was a hard doctrine, which perhaps won conviction from few philosophers. It is difficult to conceive one's domestic pet as a complex jack-in-a-box and how can one be sure that the animal does not have a view which is the opposite of the philosopher's? Vitalism – the exemption of living beings from the mechanism of inorganic Nature – has always been a strong force. Moreover, strict mechanism appeared to conflict with religion on numerous issues. Christian faith clearly imposes belief in other spiritual beings than humans: angels are involved on the best authority in the Annunciation and Nativity, and demons (whose lord is Satan) were cast out by the Redeemer himself. When the Testaments speak of spiritual beings, were they to be allowed merely an allegorical use of language or supposed to be conveying the follies of their human scribes? Again, if there were no Satan with his legions it seemed necessary to admit (with Leibniz) that God could not help introducing misery and evil into his creation. How, too, were providential actions to be understood within a mechanistic universe? Leibniz classified them as miracles, events running contrary to the order of Nature, whereas for others (Newton chief among them) the divine governance of the universe, and the divine providence, are aspects of the totality of Nature. In Newton's reluctance to exclude discourse of God from natural philosophy the influence of the Cambridge Platonists may perhaps be discerned.

For these philosophers rejected even a divinely willed mechanism as a sufficient explanation of Nature. In their view, it was counter to religion to imagine that God had thus exiled himself from the world, and an impious presumption in philosophy to suppose that it could invent devices adequate to avoid invocation of the divine guardianship and direction. Such a revolution in philosophy was unwelcome to the Cambridge men who, indeed, show little awareness of the modern notion of the progressive evolution of science and civilization, within which the separation of theology from science is an essential feature. So far from welcoming the scientific and philosophical revolution occurring in their own time, they still held firm to the belief, so crucial to the Renaissance, that the ancients in the youth of the world had possessed greater talents than they themselves and sharper eyes for perceiving the secrets of Nature, giving to the first philosophers a wisdom and understanding that 2,000 years of subsequent history had dulled rather than enriched. Newton himself knew and in part adopted this perception of antiquity; his expression of it was transferred by David Gregory to a textbook of modern astronomy in 1702. To the Platonists of the Renaissance the rebirth of learning meant an attempt to reconstruct the practices and knowledge of the ancient sages (*prisci theologi*), a conscious effort to return from the decadence of the world to the intellectual vigour of its first generations.

This 'historical' tradition had been developed in the interests of Plato against Aristotle, Plato's pupil. After the (largely imaginary and therefore anonymous) sages of Phoenicia, Egypt and Israel who were postulated as the pioneers of mathematics and philosophy, it named such figures as Thales, Pythagoras and Plato as their students and the first philosophers of Greece. Exponents of the Hermetic or magical philosophy added other names such as those of the Trismegistus and Zoroaster. Basing their assertion on a passage in the geographer Strabo, many believed that one Moschus, a Phoenician, was the first atomist: Henry More's friend Ralph Cudworth, fourteenth Master of Christ's College, was one who repeated this story, and Isaac Newton was another. Equally confidently scholars also asserted that this Moschus was to be identified with Moses and that the Jews had been the first fathers of knowledge. As late as 1691 the mathematician John Wallis (1616–1703) wrote in a sermon on the Trinity: "Tis well-known (to those conversant in such Studies) that much of the Heathen Learning (their Philosophy, Theology and Mythology) was borrowed from the Jews; though much disguised and sometimes ridiculed by them.'[1] Wallis goes on to explain that much of the evidence for this transmission and for the content of the most ancient philosophy was to be found in literary myths and symbols: so the two-faced Janus of the Romans derives 'from Noah's looking backward and forward to the World before and since the Flood' (!). Here the analogy with the mystery and symbolism of Hermeticism is obvious. This confirmation from pagan sources of the antiquity of Hebrew wisdom, manifest in the first place in the Bible, induced many scholars (Henry More among them) to study the Jewish cabala, another great fount of esoteric wisdom and historical confusion. Perhaps even more surprising (to modern readers) is the confidence of such scholars that this ancient tradition of philosophy embraced atomism, an atomism innocent of atheistic tendencies. In the earlier stages of Renaissance Platonism it seemed that the Church as a whole might prove as eclectic as the Cathedral of Siena, that this 'historical' theory of the origin of ancient Greek philosophy among the Hebrews, together with the supposedly favourable examples of such Fathers of the Church as Augustine, Origen and Clement of Alexandria in assimilating Platonism to Christianity, fully justified the close examination by modern Christians of the pagan philosophy, and their dissemination of its doctrines. At the beginning of the seventeenth century, however, Rome made manifest its disapproval of Hermetic philosophy, or at any rate of specific tenets within it, by the burning of Giordano Bruno and the imprisonment of Tommaso Campanella. In his own thought Henry More was to follow an opposite course.

Did Renaissance Platonism have a profound effect upon the Scientific Revolution of the seventeenth century? It would be easy enough to present argument and evidence to show that it gave no strong impetus to that Revolution, which in so far as its foundations were empiricist and 'positivist' was in reaction against neo-Platonism. Such exponents of the new mathematical sciences as Johannes Kepler and Marin Mersenne have left energetic criticism of the fanciful mathematical symbolisms of such Platonists as Dee and Fludd. But the juxtaposition of nascent modern science and the contemporary Platonic tradition is a complex topic in historiography, and it would be misleading to classify some ideas or programmes as 'progressive', others as 'outmoded'. Sarton on the one side can be balanced by Koyré on the other. If the legacy of Paracelsus or Cardano seems from some aspects redolent of past absurdity, treated from other points of view it may appear a force making for change and development. What of astrology and alchemy? Few would now teach that these pseudosciences were mere dead vines, bearing no fruit, and some historians claim that they made important contributions to the techniques and ideas of modern science. What of magic, so strongly linked by other recent writers with the evolution of experimental science? Probably there will never be final agreement upon the answers to such questions.

Among the oldest methodological interpretations of the changes in thought and activity that created modern science in Europe is that which emphasizes observation and experiment as the core of scientific investigation. This empiricist-inductivist tradition has above all appealed to British historians of science, among them the Cambridge philosopher and mathematician William Whewell (1794–1866), who regarded Francis Bacon as the paradigmatic prophet of modern science. Not only had Bacon advocated the fertile method of induction whereby general propositions (at the highest level, 'Laws of Nature') are founded upon the analysis of numerous instances, and consequently advocated also the necessity for the careful compilation and testing of the actual instances of phenomena to be found in Nature, but he had also foreseen the large-scale and collaborative character of this enterprise which (in Bacon's view) deserved to become a principal activity of society. Bacon held that all previous attempts by philosophers to frame a logic, ethic and science of Nature had failed. They had yielded only 'shadows thrown by words', 'mixtures of religion and science', or 'a few commonplace observations and notorious experiments tricked out to make a composition more fanciful than a stage-play'. Aristotelians and Platonists, magicians and alchemists were all following false trails (Rossi 1968: 39). Nor were the mathematical scientists of his own or former times any more successful.

Despite the unstinted praise accorded to Bacon's writings in the seventeenth century and since – not only by the British – criticism has not been lacking either. Philosophers, historians and scientists have voiced scepticism concerning the value of Bacon's nonexperiential picture of scientific research and thinking, whether considered as a programme for future endeavours or regarded as the basis for an account of what post-Baconian scientists have actually done. Some, such as Leibniz, have found his claim to furnish a logic of scientific discovery absurd; others, such as Liebig, have ridiculed Bacon for his utter ignorance of the real business of the experimental scientist. Paolo Rossi has written that Bacon's *Sylva silvarum* – an incoherent natural history highly esteemed in the seventeenth century –

is 'no different from the magical texts of Della Porta and Cardano or those of the seventeenth-century English hermetics and magicians John Dee and Robert Fludd. This is probably why Bacon's logic has been seen by so many scholars as a failure' (1968: 219). One of these was Alexandre Koyré, who poured scorn on the claim that it was Bacon who had called the scientific wits together (Hall 1987: 225–33).

But the question of the importance of systematic observation and experiment in the Scientific Revolution does not depend solely upon the merits or demerits of Bacon's philosophy, nor upon the extent of its influence; nor does the tinge of alchemy or magic associated with certain experimenters vitiate their endeavours. Koyré, following his strongly idealistic bent, emphasized the importance of the 'mathematization of Nature' (an expression meaning different things in different contexts) and hence of the Platonic philosophy that encouraged this process. A more recent trend – exemplified by the studies of Galileo published by Stillman Drake and other scholars – rejecting Koyré's denunciation of empirical discovery, reduces the importance of Platonist ideas, as of course the more profound recent work on Aristotelian philosophy in the sixteenth and early seventeenth centuries has also tended to do.[2] While no one will deny that achievement in the seventeenth century was most striking in the mathematical sciences, especially the development of mechanics that culminated in Newton's *Mathematical Principles of Natural Philosophy*, progress in observation (using telescope, microscope and other new instruments) greatly modified people's view of the universe. In some partially mathematical sciences such as astronomy, optics, pneumatics and acoustics, the role of factual discovery was crucial. Newton's new theory of light and colour sprang from 'the oddest if not the most considerable detection which hath hitherto been made in the operations of Nature' (Newton 1959, I: 82–3). Galileo's enlargement of human vision within the universe by his telescope was no less important than Copernicus's reordering of the celestial circles (see, for example, Van Helden 1985). In the Scientific Revolution new discoveries did not necessarily generate new thoughts, but they predisposed philosophers towards them.[3]

Any elements in the achievement of the Scientific Revolution attributable (directly or indirectly) to successful exploration of the phenomena of Earth and sky clearly owe nothing to Plato's teaching. The claims for the positive influence of Platonism must be restricted to its conceptual changes, whether these be regarded as paramount or not, and especially to those taking place in the mathematical sciences. Such conceptual changes may be classified as being metaphysical, epistemological, or ontological. Modern historians of philosophy and science have drawn attention to alterations in the assessment of sense-data (as when Copernicus chose to attribute three motions to the Earth, though his senses told him that it was fixed at the centre of the heavens), to shifts in metaphysical position (such as produced Descartes's deduction of the infallibility of properly conducted scientific reasoning from the existence of God), and most significantly here, the so-called 'mathematization of Nature'. We may here take this expression to denote a double assertion: (1) that a mathematical argument in natural philosophy is superior to verbal or rhetorical discussion in that its logical structure is irrefutable; (2) that mathematical propositions about natural phenomena are not mere human rationalizations of experience but actually correspond to the true structure of things. Since the book of Nature is writ, as Galileo put it, in mathematical characters, only the competent mathematician is capable of reading what God set down for humanity to discover. Though this doctrine has seemed to some modern analysts to present a rather crude reification of mathematical concepts, it was of great value to the seventeenth century in assuring mathematical philosophers that their new theories were not mere arbitrary models of reality, an assurance cognate to their confidence that such unobservables as atoms were not fictions, but really existed.

A philosopher who believes that we must judge of the reality of Nature not as the Platonist does by searching for its mathematical relations, but by consideration of what is directly made known to us, is often labelled an Aristotelian. How uncertain such characterizations may be! In the same year (1940) Randall wrote that Galileo 'in method and philosophy if not in physics ... remained a typical Paduan Aristotelian', while Koyré 'called Galileo a Platonist. And I believe that nobody will doubt that he is one' (Wiener and Noland 1957: 145; 172). Koyré made it clear that he was here identifying the Platonist and Pythagorean schools, as many did in Galileo's time (including Bonaventura Cavalieri, Galileo's pupil, whom Koyré quotes); and that the defining feature of this joint school of philosophy was the use of mathematics. But such a loose concept of Platonism both includes and excludes too much. In the first place, it excludes Platonism as Ficino and his readers understood it, not to say the whole neo-Platonic school of antiquity. Secondly, it ignores the profound difference in type between

Galileo and Kepler as students of Nature; both mathematicians, both Copernicans, yet how differing in philosophy, how little in agreement with each other! To say that both were Platonists only obscures differences in intellect and method that the historian should seek to define and understand. And again, there is a still deeper distinction to be drawn between Dee and Galileo, or Fludd and Kepler, though all these were mathematicians and Platonists. Dee and Fludd were 'spiritual mathematicians', that is, 'Renaissance Pythagoreans who sought a key to the Creation in mysticism and numerical analysis' (Debus 1978: 19). Their readers are promised 'the understanding of great mysteries', but they become neither masters of new mathematical theorems nor acquainted with verifiable numerical relations within the realm of experience. As Debus and others have pointed out, the superficial likeness between Dee's notion of the monad (or Fludd's cosmic symbolism of geometric figures) and Kepler's cosmology of the Platonic bodies is quite misleading, for Kepler (unlike them) used his Platonist cosmology as a springboard from which to investigate testable relationships within the body of quantitative data accumulated by astronomers (Debus 1978: 123–6). Kepler uses the Platonic *a priori* – as many mathematical philosophers have done – to make an entrance into the *a posteriori* of technical astronomy; Dee and Fludd never desert its fancies:

> one of the contrasts Kepler points out between his own work and that of Fludd is that Fludd's harmonies ignore actual units [of measurement] and use abstract numerical relationships, whereas Kepler [in *Harmonices Mundi*] finds musical ratios among quantities measured in the same units, such as the extreme angular speeds of planets as seen from the Sun. (Field 1988: 184–5)

Fludd made clear (in his replies to Kepler's criticisms) his confidence that things natural and supernatural are not bound together by numbers that are only to be discovered by painstaking investigation of phenomena, but instead

> by particular formal numbers. The mystery of these occult numbers is best known to those who are most versed in this science, who attribute the Monad or unity to God the artificer, the Dyad or duality to Aqueous Matter, and then the Triad to the Form or light and soul of the universe, which they call virgin. (Field 1988: 185)

Kepler found another cause for criticism in Fludd's adhesion to the now untenable Ptolemaic universe: Fludd was a vehement anti-Copernican (Field 1988: 179–87).

It would be an elementary error to suppose that the spherical trigonometry and other branches of mathematics employed by Kepler (or, *mutatis mutandis*, Galileo) descended in any way from Plato. The origin of technical mathematics, as applied to the physical sciences, was (so far as we know) with Euclid, Apollonios, Archimedes, Ptolemy and Pappos. In a limited, technical sense Kepler might be called an 'Apollonian' astronomer, as Copernicus had been 'Ptolemaic' (and Islamic) and Galileo was 'Archimedian'. These epithets are at least as appropriate as the all-embracing 'Platonist'; Kepler, the founder of elliptical astronomy, expresses his debt to Apollonios' *Conics*, and Galileo rightly regarded Archimedes as the ancient mathematician whose example and method he himself most closely followed, Archimedes being after all the archetype of the mathematical philosopher, Plato the archetype of the mathematical metaphysician. Galileo was a practical inventor and a realist (among other attributes). He apprehended perfectly the difference in character between Archimedes' resolution of the problem of Hiero's crown and Plato's utterances about the role of geometric shape in the composition of substance in general. Kepler, too, thoroughly understood that the cash-value of planetary metaphysics is ultimately tested by the accuracy of ephemerides. Even if a somewhat nebulous distinction be attempted between physicalist Pythagoreans and metaphysicalist Platonists one may still hesitate to accept the derivation of the seventeenth-century mathematical philosophy of Nature from either ancient example. Certainly, on occasion, the seventeenth-century philosopher seems to lean upon an ancient Greek avatar *ex post facto* as a rhetorical justification for his arguments, when it would be a misinterpretation to look for a simple, direct inspiration; Copernicus's allusions to the Pythagoreans may be an example. And this appears to be the case also with Newton's late references to the Pythagoreans and their supposed familiarity with the law of gravitation: such a scholarly influence is not to be found in Newton's early notebooks (Casini 1984).

There is an obvious attraction of simplicity in the hypothesis that the humanist rediscovery of Plato in the fifteenth century prompted a reaction against Aristotle that favoured a mathematical rather than a qualitative philosophy of Nature and so gave rise to the Scientific Revolution. But as a key to a complex history this hypothesis is too simple. For, on the one hand, the reading of Plato and *a fortiori* of the neo-Platonists pointed towards many different destinations, perhaps more markedly mystical than 'scientific' in the sense that Kepler and Galileo, Descartes and Newton were 'scientists'. Nor were all Platonists spiritual mathematicians of the type of Dee or Fludd; none of the Cambridge Platonists falls into this category. For all of Henry More's sympathy for Galileo, it would be strange to apply the same label to these two men without distinction.

In the second place, we must recognize that Aristotelian philosophy did more than continue to flourish throughout the Renaissance; in the biological sciences it made important contributions to the Scientific Revolution. Despite the damage suffered by Aristotle's physics and cosmology, the teaching of his logic continued to have a fruitful formative influence upon the scientific revolutionaries. But it is even more important that among the innovative biological investigators of the sixteenth and early seventeenth centuries several (Andreas Cesalpino, Hieronymus Fabricius ab Aquapendente and William Harvey) can properly be described as neo-Aristotelians: their experimental method and their conception of the living organism came from Aristotle's Lyceum, owing nothing to Bacon's idea of 'Salomon's House'. In embryological studies, in zoology and even in physiology, Aristotle was a potent master, as was his pupil Theophrastos in the investigation of plants. In biology and medicine, as in astronomy, revolutionary changes in the forms of explanation were consistent with the maintenance of continuity in philosophy and methodology.

Reverting to the mathematical sciences proper, it is difficult to discern Platonic influences in such writers on optics as Maurolyco, Porta and Kepler, or such writers on astronomy as Copernicus, Magini, Maestlin or Tycho Brahe, or those on mechanics such as Commandino, Benedetti, Ubaldi or the youthful Galileo. Plato himself had shown little respect for the mathematical sciences based on observation, measurement and calculation, and none for the 'Vitruvians' – the surveyors, architects, engineers and mechanics, whose work (still fragmentarily visible today in tunnels, bridges, aqueducts and harbour-works, cities and temples) was recorded in writing by Greeks from Ctesibios (third century BC) onwards. One aspect of the Renaissance was the revival of scholarly interest, combined with renewed practical achievements, in the Vitruvian subjects ranging from the mathematical science of painting in perspective to the engineering of mines and fortifications.

These subjects too represent a 'mathematization of Nature' which was an historical process much wider than, as well as distinct from, the mathematization of philosophy, though the two have often been conflated. The analogy between numbers and Nature may be made manifest in a vast spectrum of ways, from the use of the surveyor's chain and plane-table and the seaman's quadrant to the metaphysical notion that God is necessarily a geometer. As E. W. Strong emphasized more than half a century ago (1934) the use of numbers in natural philosophy has no precise methodological significance, and entails no particular ontology. It embraces equally calculations of the calendar and of the table of atomic weights. Platonism and neo-Platonism define the metaphysical end of a spectrum which already in Greek antiquity included exact mathematical sciences, axiomatically constructed, making no metaphysical assertions, from which could be made verifiable/falsifiable predictions about the movements of the heavenly bodies, the properties of mirrors and light-rays, or the behaviour of floating bodies. It is true that the first practitioners of these sciences made no very exalted truth-claims; *if* the axioms they postulated held true in Nature, then the consequences drawn from them would also hold true. It was certainly possible that the axioms might not be exact: that light might not always travel in straight lines, that water might not be a perfect fluid as defined by Archimedes. In fact, Galileo (not Plato) seems to be the first philosopher to confer upon the exact, quantitative mathematical sciences, rather than upon a metaphysical philosophy of Nature dealing with numbers, the supreme status of being the highest knowledge of Nature that humans can attain. Similarly, Copernicus and Kepler were the first mathematical astronomers to give their theorems the status of metaphysical truth: they asserted that the kinematical systems of planetary motion that they described really exist in Nature.

We may plausibly believe that this transfer of the principal truth-claim as regards the relation between mathematics and Nature from philosophy of the Platonic or neo-Platonic kind to exact sciences of the Ptolemaic or Archimedian model was the foundation for the Scientific Revolution in the physical sciences, because it was this transfer that closed the gap between the mathematicians' ways of investigating Nature and those of the philosopher, a hiatus that had existed since the fourth century BC. It was now possible for mathematicians to maintain that a mathematical theory is true if its physical assumptions are confirmed by experiment, and the predictions it makes are also verified by experiment. Hence the supreme importance for the later history of science – not astronomy alone – of the work of Copernicus, Kepler and Galileo; hence also the significance of the distinction so firmly and so clearly drawn by the exponents of the new mathematical sciences of the seventeenth century (such as Kepler and Mersenne) between *their* 'mathematization of Nature' and that undertaken by adherents of the Hermetic and neo-Platonic schools like Dee, Bruno and Fludd. These we may now leave with D. P. Walker's judgement of the last named: 'Ficino's analogies between cosmic, musical and human spirits led to magic; Kepler's mathematical ones to science; ... and Fludd's were not meant to lead anywhere at all' (1958: 117).

There is a further complexity still to be considered: the Cartesian 'logical glissade' by which reasoning *more geometrico* no longer signified mathematical reasoning. Descartes had been faced by the fact that it was humanly impossible for him to extend widely through physics and cosmology the mathematical analysis so successfully practised in his *Dioptrique* and *Météores* and, in particular, there seemed to him insuperable difficulties in mathematizing the science of motion (for he rejected the progress made in this respect by Galileo). Accordingly, he argued that if a rhetorical argument be set out in the propositional form of the geometers, starting from trustworthy premises and articulated with a logic as impeccable as that of the mathematicians, it could be regarded as producing conclusions as certain as those reached by reasoning *more geometrico.* In this claim, tantamount to striking an equivalence between *rational* and *mathematical,* Descartes was followed by a number of later philosophers, among them Leibniz, Spinoza and Henry More. Implementing it, Descartes gave to his principal contribution to our understanding of the physical universe, *Principia Philosophiae* (1644), a structure that is wholly rhetorical; geometry is banished, and a mechanical imagery justified by reasoning *more geometrico* takes its place. It is interesting that More, the Platonist, did not criticize Descartes for this Aristotelian desertion of geometry. Nor did More, who before studying Descartes had absorbed the mathematical philosophy of Galileo with much satisfaction, and who from middle life onwards was a relentless critic of Descartes's anti-spiritual mechanism, ever point out the fallacy involved in equating rhetorical argument with formal logic, whether mathematical or not.

The philosophers' willingness to substitute rhetorical argument for mathematical reasoning, when convenient, is one instance of the fact that mathematics did not, and could not, provide the unique route to revolutionary developments in the sciences. Further, the case of Descartes shows how indefinite the label 'mathematical philosopher' may be when not otherwise defined. Particularly, one cannot prudently equate this term with 'Platonist'; many who applied mathematics to the understanding of Nature, like Simon Stevin and Thomas Harriot, were not Platonists; many Platonists (including Henry More) were not mathematical philosophers of the stamp of Galileo, Kepler and Isaac Newton. In the days of his *Philosophical Poems*, More made efforts to substantiate his views by geometrical reasoning, but later this appealed little to him. In 1664 he wrote to his friend Lady Conway that he was renewing his 'acquaintance with such Mathematical Theorems as I was in some measure conversant with before I fell a Theologizing, which was so long [ago] that I had almost forgott all that little I knew in Geometry' (Nicolson 1930: 231). Basically, More possessed a sound knowledge of both algebra and geometry; perhaps the revision of which he wrote was undertaken for the benefit of his pupils whom (for example) he conducted through Descartes's scientific treatises.

In general the Cambridge Platonists, like More in particular, were not mathematical philosophers of any kind, rather they were above all divines. The thrust of their teaching was to enlighten humanity's knowledge not of Nature but of God, and their great purpose was to vindicate universal mind against universal matter. To this end even the reconciliation of Platonism and Christianity was subservient.

4

The Cambridge Platonists

THE USUAL claim for the existence of a coherent and co-operative group of philosophers called the Cambridge Platonists must appear weak to anyone who applies a critical eye to the membership of the group and to their various activities. For a start, the distinctions between the early members of the supposed group (who were Emmanuel men) and the later ones (who were Christ's men) are profound. The Emmanuel men were preachers, the Christ's men were writers. The latter possessed and were influenced by a consciousness of their intellectual relationships to the new scientific movement of the seventeenth century of which the former were wholly innocent. In the sense of the word current during recent centuries, only Cudworth and More could be designated as philosophers; the Emmanuel men were theologians.

Nevertheless, the term 'Cambridge Platonists' (and variants of it) have been generally employed without question, as though the group constituted an identifiable school of philosophy. This was far from being the original manner of identifying the 'Cambridge School'. If we may follow Tulloch, it seems that the Cambridge group was first labelled, at the time, as that of the 'Latitude men', that is, as theologians unwilling to prescribe some precise doctrine of Christianity and proscribe all others as heretical and damnable. But where this label first appears, in a pamphlet of 1662, the writer already asserts that it is a mere term of abuse, 'an image of clouts, that men set up to encounter with for want of a real enemy' (Tulloch 1874: 36). Those recent scholars who have discovered that the Cambridge Platonists were Latitudinarians are, as it were, putting the cart before the horse.

Bishop Burnet, in a more extended account, linked the Cambridge men together as shining counter-examples to the general depravity and idleness of the Church of England after the Restoration. They constituted, in his eyes, 'a new set of men ... of another stamp [who] studied to assert and examine the principles of religion and morality on clear grounds, and in a philosophical method. In this way More led the way to many that came after him' (1874: 32–4). Burnet thus implicitly ignored all manifestations of 'Cambridge Platonism' during the ascendancy of the Puritans, and its development as a reaction against that ascendancy. The apparent inadequacy of his chronology is, in a measure, vindicated by dates of publication: the works of the Emmanuel men were all posthumously published and only Culverwel could rival More in priority of publication. However, his assertion that the Cambridge scholars 'declared against superstition' must be taken in a special sense: they opposed ritual superstition in Divine Service. Burnet further hints at that common attribute of the Cambridge men so much emphasized by later historians: their rationalism. So he wrote of Benjamin Whichcote:

> He was much for liberty of conscience; and being disgusted with the dry systematical way of those times, he studied to raise those who conversed with him to a nobler set of thoughts, and to consider religion as a seed of a deiform nature, to use one of his own phrases. In order to [achieve] this, he set young students much on reading the ancient philosophers, chiefly Plato, Tully and Plotin, and on considering the Christian religion as a doctrine sent from God, both to elevate and sweeten human nature. (1874: 32)

Other contemporary observations of a link between the Emmanuel theologians and the reading of Plato and the Platonists are as vague as this, and relate chiefly to Whichcote.

It is evident that Burnet was concerned with the influence of these Cambridge teachers upon the history of religion in England, and so likewise were the historians of two or three generations ago such as Tulloch and Powicke. Campagnac's selections (1901) from *The Cambridge Platonists* were also wholly theological in character and included nothing from the Christ's men, More and Cudworth. The Cambridge group was recognized as one of Platonist (we might say, after More, Platonizing) *theologians* rather than as adherents to a continuous Platonic or neo-Platonic tradition of philosophy. Even More and Cudworth were treated – not unjustifiably – from a purely theological standpoint. The historical treatment of the Cambridge men as a school of Platonist *philosophers* seems to be an innovation of the twentieth century, given countenance by the great authority of Ernst Cassirer (1932; English translation, 1953). Even Cassirer had very great difficulty in establishing any point of contact between Renaissance Platonism and 'Cambridge Platonism'; indeed, it seems that in terms of literary influence there is virtually none. The Christ's men studied such authors as Marsilio Ficino and Cardano, but their Emmanuel predecessors had not done so. It is not even clear how Whichcote enabled his pupils to read 'Plato and his scholars' since none of the relevant works had yet been printed in England and no copies of the continental editions are recorded as being present in Cambridge libraries in Whichcote's time.

Historians of Cambridge Platonism have expressed its central doctrines in vague and doubtful terms. The older ones emphasized, rightly, that members of the group were divines first, philosophers second (if at all). Thus F. J. Powicke: '[Neo-Platonism] did not mould the substance, or even the forms, of their thoughts, to any great extent. They drew far more from the Bible; and their acknowledged Master was Christ' (1926: 21). In his interpretation, the pagan tradition informed and strengthened beliefs which the Cambridge men had already formed from Christian sources: 'Whichcote and his fellows certainly read Plato, but they read Plotinus far more, and found in him a welcome confirmation of some of their leading ideas, though they did not derive them from him' (1926: 18).

A generation later, Ernst Cassirer laid less emphasis on the Christian roots of Cambridge Platonism, but agreed with Powicke in finding the source of its opposition to materialism in Plotinus, whose doctrine Cassirer paraphrases thus: 'It is impossible that a chance concurrence of bodies should bring about life, and that the non-mental should bring about mind' (1953: 136). In part, at least, Cassirer applauded Powicke's judgement that at bottom it was not the metaphysics and still less the natural philosophy of neo-Platonism that won adherents at Cambridge, but 'its religious spirit and its ethical idealism' (1953: 41). Cassirer declared succinctly that 'The *apriori* of pure morality was the starting-point of the Cambridge Platonists' doctrine': it furnished their defence against empiricism and atheism (1953: 57). All writers emphasize the rationalism and broad theological compass taught at Cambridge; none attributes to Cambridge a fondness for the esoteric and the magical trends condemned by Augustine (with the exception of Henry More). None attributes to it any scholarly achievement in the study of Plato and his followers; rather it is admitted that its use of the ancient writings was highly indiscriminate. The claim that the Cambridge men were Platonists seems, in the end, to reduce to the fact that they were idealists opposing materialists, rationalists opposing dogmatists and intuitionists opposing sensationalists.

The composition of the Cambridge group is as indefinite as its intellectual coherence. Burnet named Benjamin Whichcote, Ralph Cudworth, John Wilkins, Henry More and John Worthington as the chief of the Cambridge divines. Tulloch, the modern pioneer of studies relating to them, listed in addition John Smith as a major figure and added to the minor ones the names of Nathanael Culverwel, George Rust, Edward Fowler, Simon Patrick, Joseph Glanvill, John Norris and Sir Thomas Browne. Thanks to the work of Campagnac and Powicke it may now be said that Smith and Culverwel have long been established as major Cambridge Platonists, while Powicke further added to the group the name of Peter Sterry.

Of all these, we may at once dismiss the names of John Wilkins, Joseph Glanvill, John Norris and Sir Thomas Browne, for none had any association with the University of Cambridge (save that Wilkins was Master of Trinity College for a short space in 1659). Whether or not they were Platonists, whether or not they had close links with the Cambridge men – as is true of Glanvill – they cannot possibly be described as 'Cambridge men' themselves. The remaining minor figures pass the test of academic affiliation, but none exercised a powerful influence within the University, or exercised it through means that can in any way be described as philosophical. Edward Fowler (1632–1714), who became Bishop of Gloucester, was certainly a Latitudinarian. George Rust (d.1670) became an Irish bishop; he had been a Fellow of Christ's College (1649–59) and was a pupil and friend of Henry More. Simon Patrick (1626–1707), Bishop successively of Chichester and Ely, is a more weighty historical personage: he was at Queens' College, Cambridge, with Culverwel, whose funeral oration he delivered. He, too, is numbered among the reformers of the Caroline Church and was a prolific author. John Worthington (1618–71), Master of Jesus College, Cambridge, from 1650 to the Restoration, was a product of Emmanuel and published the works of the distinguished tutor of Christ's, Joseph Mede (1586–1638). He too was a close friend of Henry More, and by his correspondence with literary men, rather than original literary work of his own, was closely involved with the lively intellectual currents of his time. But again no place in the history of English philosophy can be assigned to Worthington. As for Peter Sterry (d.1672), Oliver Cromwell's chaplain, who is described as a Platonist divine, he departed early from Cambridge for the Westminster Assembly and a London career. None of these men will figure in what follows save incidentally.

Essentially, then, the group of the Cambridge Platonists who are serious figures in English intellectual history numbers five: Whichcote, Smith, Culverwel, More and Cudworth, of whom only the two last can be reckoned philosophers.

Benjamin Whichcote (1609–83), a Fellow of Emmanuel (1633), became a popular Cambridge preacher, his background being Puritan and his wife the widow of the first Governor of the Massachusetts Bay Company. Appointed in 1644 Provost of King's College, he was chosen Vice-Chancellor in 1650–1, but was forced to retire to London at the Restoration. There he settled at the Church of St Lawrence Jewry. His writings, of which his *Aphorisms* and his sermons are the most important, were wholly devoted to Christian belief and doctrine. Whichcote was a man of toleration, whose abandonment of Calvinist orthodoxy did not go uncriticized in his lifetime. He was accused, one might judge unjustly, of reading Plato and his scholars before the Scriptures and of giving reason too high a place in relation to the mysteries of faith. One of Whichcote's *Aphorisms* reads: 'Nothing *without* Reason is to be *proposed*; nothing *against* Reason is to be *believed*: Scripture is to be taken in a rational sense' (Campagnac 1901: 70). And elsewhere he wrote:

> Man is not at all settled or confirmed in his Religion, until his Religion is the self-same with the Reason of his Mind: that when he thinks he speaks Reason, he speaks Religion; or when he speaks religiously, he speaks reasonably; and his Religion and Reason is [*sic*] mingled together; they pass into one Principle; they are no more two, but one: just as the light in the Air makes one illuminated Sphere; so Reason and Religion in the Subject, are one Principle. (1901: 55)

Whichcote's detestation of dogmatism may be illustrated by the sentence: 'Determinations beyond Scripture have indeed enlarged Faith; but lessened Charity and multiplied Divisions' (1901: 71). Freedom of belief he grounded on the great Reformers' assertion of the rights of individuals to build their own faith: 'I will not make a Religion for God: nor suffer any to make a Religion for me' (1901: 12). Cassirer quoted from Whichcote's *Sermons* the saying: 'Nothing is desperate in the condition of good men; they will not live and die in any dangerous error', meaning of course that God accepts all righteous worship and moral living (1953: 124).

Whichcote seems seldom to allude in his writings to Plato and the Platonists, and certainly it was contrary to his style to make any parade of Greek learning. A Platonic influence that is general rather than particular, moral rather than intellectual, may be detected in such of his aphorisms as 'Truth is connatural to a man's soul ... It is the chiefest of Good Things for a Man to be *Himself*' (Campagnac 1901: 68). In this respect Whichcote was not followed by his junior fellow collegian (and possible pupil) Nathanael Culverwel (?1618–51). His elaborate prose style, more familiar perhaps to modern readers as coming from the pen of Sir Thomas Browne of Norwich, was much praised by the older historians. This style, and Culverwel's elaboration of Whichcote's rationalism, may be sampled in the following passage from the opening pages of his *Discourse of the Light of Nature*:

> But some are so strangely prejudiced against *Reason* (and that upon sufficient reason too, as they think, which yet involves a flat contradiction) as that they look upon it not as *the Candle of the Lord*, but as on some blazing Comet, that portends present ruine to the Church, and to the soul, and carries a fatal and venemous influence along with it. And because the unruly head of *Socinus*, and his followers, by their meer pretenses to *Reason*, have made Shipwreck of *Faith*, and have been very injurious to the Gospel; therefore these weak and staggering apprehensions are afraid of understanding any thing; and think, that the very name of *Reason*, especially in a *Pulpit*, in matters of Religion, must needs have at least a thousand Heresies couched in it. If you do but offer a *Syllogism*, they'd straightaway cry it down for *carnal reasoning*. What would these men have? Would they be banished from their own *essences*? Would they forfeit and renounce their understandings? or have they any to forfeit, or disclaim? (1901: 214)

Culverwel constantly intruded Greek words and phrases into his English prose, and almost as frequently appealed to the authority of Plato: 'But Plato, who was more spiritual in his philosophy ... But that *Divine Philosopher* do's most admirably discover ... that *speculative* Lawgiver, *Plato* I mean ...'. To be fair, it should be added that Culverwel quotes very frequently also from Aristotle, from the Fathers of the Church, the 'Schoolmen' and even 'Fryer Bacon'.

The human intellect he calls, from its power to shed light into the darkness of ignorance, *the Candle of the Lord*. What truths does humanity perceive in its light?

> First, That all the Moral Law is founded in natural, and common light of *Reason*.
> Secondly, That there's nothing in the mysteries of the Gospel contrary to the light of *Reason*; nothing repugnant to this light, that shines from *the Candle of the Lord*. (1901: 221)

The problem about such appeals to reason, or rather about the confident assertion of the necessary conformity of Reason and Faith put forward by the Cambridge Platonists, is the ambiguity in the word *reason* already exploited to rhetorical effect by Culverwel in the long passage just quoted. The strength of the argument that the Christian faith is rational rests on the presupposition that reason is common to all, Christian and pagan alike. The specific claim is that if any individuals will employ their native reason honestly, without prejudice, in the most humanly natural way, they will find the Christian faith to be true. If reason were not to be regarded as objectively clear and compelling for all people, as it is among mathematicians, then it would have to be admitted that the Mohammedan may subjectively as well regard the faith of Islam as rational; that the Catholic may appeal to reason against the Protestant with the same force of argument as the Protestant against the Catholic. And if reason is to be regarded (in matters of philosophy and theology) as subjective rather than objective (as it is in mathematics) then any assertion of the conformity of the Anglican Christian faith with reason becomes meaningless; any other sect may make an identical claim.

Now it is obvious that if the word 'reason' be taken to mean 'the process of ratiocination', or more narrowly still 'the use of a logical calculus', then in principle (if not in historical fact) all may agree on the nature of that process or calculus, and again in principle (if not in fact) this may be put into a universal symbolic form. But reason is not considered simply in this way as the instrument of thought. It is also applied to the very material upon which thought operates, most vaguely when the lawyer appeals to the concept of what seems reasonable to 'the man on the Clapham omnibus'. It is impossible to *reason* with anyone – in order to bring about a change of opinion or conduct – by invoking a logical calculus: it is necessary to appeal to common truths, experience, probabilities, in other words to a body of information supposedly correct as to fact. The calculus is relevant only in drawing conclusions from this information. *Reasoning* may consist in the demonstration that one contingent proposition is more likely to be valid than another: that oil prices are more likely to rise than to fall, or that the dodo is more likely to have existed in the historical past than the dragon. Such contingent propositions cannot possibly be verified by the logical calculus – the calculus is only useful when applied to relevant economic and political facts in the one case, to the narratives of mariners and chroniclers in the other. So, in matters of religion, ratiocination cannot remove the necessity, in the last resort, for the declaration 'I believe the words of St Mark to be true' or 'I believe the words of Mahomet to be true.'

Similarly, no logical calculus can determine whether the natural world really exists or not, without reference to the evidence of our senses, still less whether a supernatural world also exists (as Henry More believed) or does not exist (as Thomas Hobbes believed). In recent times the term *rationalist* has commonly been taken to imply belief in the natural world and denial of the supernatural (cf. Brinton 1950: 334).

Perhaps one might make the same point in a different way by observing that the assertion of the Cambridge Platonists (or for that matter of the authors of the *Bridgwater Treatises* one and a half centuries later) that Anglican Christianity can and must be justified by the exercise of human reason is itself a proposition that cannot be demonstrated by any logical calculus without the previous supply of relevant information. This is not to say that the claims for rationality made by the Cambridge Platonists were empty of meaning, because they are far from demonstrable or even clear. Not at all. They were ethically significant because they promoted tolerance in religion; Whichcote, More and Cudworth all recognized implicitly that different formulations of Christianity appear reasonable to different individuals, and held that each should stick to his own, according to his nature. Psychologically the rationality of the Cambridge Platonists encouraged individuality – though all of them, it seems, distrusted religious 'enthusiasm'; as Whichcote wrote: 'Among Christians, those that pretend to be *Inspired*, seem to be Mad; among the Turks, those that are Mad, are thought to be Inspired' (Campagnac 1901: 75). Historically, their rationality favoured impartiality and scholarly examination of the Christian history and tradition.

It should not be forgotten that, taken in the transcendental sense favoured by Henry More and cognate thinkers, reason furnishes another argument for the existence of God and the immortality of souls. For if there is reason in the universe, if things have any cause and purpose, then there must be a God who embodies and expresses reason: for God is not to be thought of as an arbitrary despot. As More wrote in one of his *Philosophical Poems* (1647):

> If God do all things at his pleasure,
> Because he will, and not because it's good,
> So that his actions will have no set measure,
> Is't possible it should be understood
> What he intends?
>
> (Lovejoy 1948: 188)

Just as reason in the universe implies a Divine Reasoner – and by the same token design in the universe implies a Divine Designer – so also reason within the creation entails the Creator's having a purpose, which is the good of his creation; how indeed (More argues) could we be sure of salvation if we 'thus pervert / The laws of God, and rashly do assert / That will rules God, but God rules not God's will'. Twenty years later, in his *Divine Dialogues*, More repeats the same sentiment: 'the belief of divine Providence and that all things are carry'd on for the best at the long view which I profess I am as fully persuaded of as I am of any thing in Mathematicks' (Lichtenstein 1962: 29).

Leibniz was to strengthen this line of thought by the argument that if God does *not* purposefully seek good for his creation then his will must be governed by chance, which is the same as to say that there is no God in the universe. Thus, in short, if (contrary to Hobbes) we posit that the universe is not ruled by chance but by reason, then we must also believe in the existence of a God who pursues the good of the universe.

Henry More, like other subtle theologians, rejected the crude view popular in the hymnals that God created the universe for the sake of the chorus of praise He would receive from it. On the contrary, the creation was for its own sake and God 'takes Pleasure that all his Creatures enjoy themselves that have Life and Sense, and are capable of any enjoyment' (*Antidote against Atheism*, Lovejoy 1948: 188): 'All what he doth is for the creature's gain, / Nought seeking from us for his own content' (1948: 351–2). From this consideration More argued further that, in contemplating the rational creature Man through eternal ages, God would have less satisfaction if humans were mortal than if their souls were immortal: 'there is nothing considerable in the creation, if the rational creature be mortal.' Therefore God must have made the human soul immortal.

However, as Powicke pointed out long ago (1926: 136–49), the metaphysical rationalism and doctrinal tolerance of the Cambridge Platonists did not entail indifference in belief, and the inherent limitations to their rationalism left scope for a residual Puritanism – most strongly evident in Culverwel's sermons – that was inevitable in Republican Cambridge. The candle of the Lord was, after all, but a feeble glim compared with the effulgent splendour of the Redeemer himself, and this was to be found shining in the pages of Scripture. If Socrates was more pleasing to God than Aristophanes, or Augustus than Tiberius Caesar, and the divine punishment of the one therefore more lenient than that of the other (as Culverwel somewhat quaintly conceives) it was due to God's own will that it was so, and His grace measured out accordingly to these pagans. Culverwel wrote of a faith that was in conformity with the laws of Nature: but where (in his view) does Nature speak? It is in the Bible: 'For this is the voice of Nature itself, that whatever God reveals must needs be true; and this common principle is the bottom and foundation of all faith to build upon' (1926: 143). Reasoning begins with the assurance that Scripture is the indubitable word of God. Reason is therefore a 'derivative light'.

Culverwel, however, is hardly a Platonist at all – or so a reader might conclude from the eleventh chapter of his *Discourse on the Light of Nature*. He gives just and measured praise to Aristotle in antiquity and Bacon in modern times. He is aware of the recent 'brave *improvements* made in *Architecture*, in *Manufactures*, in *Printing*, in the *Pyxis Nautica* [mariner's compass]', and accepts Harvey's discovery of the 'constant *Circulation of Blood* through all our *Veins* and *Arteries*': 'For here's no limiting and restraining men to *Antiquity*, no *chaining* them to *old Authors*, no *regulating* them to I know not what *prescribed Forms* and *Canons*.' Culverwel condemns the Church of Rome for its treatment of Galileo's mathematical demonstrations and use of the telescope, Galileo's reward being 'to rot in an *Inquisition* for such *unlicenced Inventions*, for such *venturous undertakings* (Campagnac 1901: 127; 306–7). Little sign of the *prisci theologi* here! Culverwel criticizes Origen for following Plato in his contempt for the human body, a contempt entailing the error that the *'Souls* of men were long extant before they were born' and the conceit of innate ideas carried with the perinatal soul. Not even Locke could be more opposed to innate ideas than Culverwel: 'I wish, that the *Platonists* would but once determine, whether a *Blind Man* be a *competent Judge* of *Colours* by virtue of his *connate species*; and whether by supply of these *Ideas*, a Deaf *Man* may have the *true notion* of *Musick*, and *Harmony*?' (1901: 292). And Culverwel goes on to praise Aristotle for purging such Platonic 'dross'.

John Smith (1616–52), so close a contemporary of Culverwel, entered Emmanuel three years after him, later being moved by the Earl of Manchester to Queens' College. He is reckoned the most intellectually brilliant of the Emmanuel Platonists and was still read late into the nineteenth century. Coleridge was among the admirers of his writings. Other than Henry More, Smith alone among the Cambridge Platonists is to the slightest degree associated with mathematics. He is the only one among them to make frequent reference to Plotinus as well as Plato, though no more inclined to magic than his fellows, nor less of a theologian. Worthington was again the editor of this young scholar's works, published as *Select Discourses* in 1660, works which, Tulloch judged (a century and more ago), 'no spiritually thoughtful mind can read unmoved'.

Like Culverwel, Smith was fond of inserting snippets of Greek into his (rather simpler) prose. One may well feel with Tulloch that 'The Cambridge Platonists carried the system of quotation to great excess. They leant too fondly on the past, and made too much of ancient wisdom' (Tulloch 1874: 137). In the following passage Smith explores the idea, dear to this Cambridge school of theology, that 'Divine truth is better understood as it unfolds itself in the purity of men's hearts and lives, then in all those subtil Niceties into which curious Wits may lay it forth.' He begins by discussing Jesus's example, and quoting St Peter, then continues (the Greek here omitted):

> Neither was the Antient Philosophy unacquainted with this Way and Method of attaining to the knowledge of Divine things: and therefore *Aristotle* himself thought a young man unfit to meddle with the grave precepts of Morality, ... And it is observed of *Pythagoras*, that he had several waies to try the capacity of his Scholars, and to prove the *sedateness* and and *Moral* temper of their minds, before he would entrust them with the sublimer Mysteries of his Philosophy. The *Platonists* were herein so wary and solicitous, that they thought the Mindes of men could never be sufficiently purg'd from those earthly dregs of Sense and Passion, in which they were so much steep'd, before they could be capable of this divine *Metaphysics*: and therefore they so much solicite *a separation from the Body* in all those that would, as Socrates speaks, sincerely understand Divine Truth; for that was the scope of their Philosophy. This was also intimated by them in their defining Philosophy to be *a Meditation of Death*; aiming herein at only a *Moral* way of dying, by loosening the Soul from the Body and this sensitive life ... and therefore besides those *aretai kathartikai* (purifying virtues) by which the Souls of men were to be separated from sensuality and purged from fleshly filth, they devised a further way of Separation more accommodated to the condition of Philosophers, which was their *Mathemata*, or Mathematical Contemplations, whereby the Souls of men might further shake off their dependency upon Sense ... Besides many other waies they had, whereby to rise out of this dark Body ... several steps and ascents out of this miry cave of mortality, [an allusion to the well-known allegory in Plato's *Republic*] before they could set any sure footing with their Intellectual part in the land of Light and Immortal Being. (Campagnac 1901: 86–7)

If any hint of chemical or magical techniques for enabling the soul (supposedly) to leave the body, well documented in other cultures, may perhaps be discerned in this passage, John Smith was too innocent or discreet to explore them.

The scholarship, the Greek, the doctrines of ancient philosophers, are after all incidental to Smith's Christian purpose, as they are with all the earlier Cambridge Platonists. The Christian theologian can discover no fresh truth in pagan philosophy; rather, he confirms from it the truth of which he is already apprised from divinely inspired sources. He is convinced that the inkling of truth he finds in the pagans is proportionate to the illumination which God vouchsafed them. In Powicke's phrase: 'the ancients are used to illustrate what [Smith] is already sure of ... assurance has its foundation in experience supported by Scripture' (1926: 108). Some recent discussion of the citation of the ancients by seventeenth-century authors, especially their citations from the pre-Socratics and the Platonists, has perhaps insufficiently weighed such uses of the texts. In this century especially, consideration of an ancient text did not necessarily imply that it was being invoked as the prime authority for the point in question, still less should it be taken to imply that the text quoted inspired the author's own view of the matter. The Platonizing theologians of Cambridge certainly supposed that their own understanding of divine truth and of the relations between God and humanity was superior to those prevailing among the pagans in antiquity, just as Copernicus had understood that his apparatus of heliocentric astronomy was superior to that of the ancient Pythagoreans. There could in the seventeenth century be no effective contest for metaphysical eminence between 'Pagans and Christians' to parallel the dispute between 'Ancients and Moderns' and indeed, so far as the natural sciences and medicine were concerned, in the late seventeenth century the fact that overwhelming advances had been effected in the last couple of generations was virtually beyond dispute.

The last of the Cambridge men to be briefly considered at this point is Ralph Cudworth (1617–88), also educated at Emmanuel under Whichcote, where he became a successful tutor; possibly John Smith was his pupil, as Sir William Temple certainly was. His earliest writings – and Cudworth was an assiduous penman, though slow to publish in later life – already demonstrate his erudition. In Tulloch's words: 'His own thought is buried amidst the mass of philosophical antiquarianism in which he sets it', including, among Renaissance Platonists, Pico della Mirandola. Cudworth possessed, at the time of his death, a very large library of some 2,500 volumes; his interest in the science of his day is evident from the presence in it of large numbers of scientific and medical books, 'including almost complete runs of works by Brahe, Kepler, Descartes, Gassendi, Bacon and Boyle as well as additional works by Lansberg, Briggs, Mydorge, Boulliau, Harriot, Harvey, Torricelli, Wallis, Spinoza and [Isaac] Newton' (Feingold 1984: 52). However, despite the title of his principal work, *The True Intellectual System of the Universe* (1678), Cudworth – like all the Cambridge

Platonists – was a religious philosopher. His first book took *The Lord's Supper* as its subject. He was sufficiently in tune with the theological spirit of the Commonwealth to be promoted Master of Clare Hall (1644) and in the same year Regius Professor of Hebrew, justified by his special interest in Jewish history and theology. Ten years later, along with Whichcote and Worthington, Cudworth contributed to the celebratory volume *Oliva Pacis* presented to Cromwell; he also shared in the similar salute from the University to Charles II in 1660. Also in 1654 Cudworth transferred to the Mastership of Christ's College; tradition has it that Henry More declined the Mastership on this occasion. Despite his involvement in public affairs during the Protectorate of Cromwell, Cudworth survived the Restoration and the new disposition of the Anglican Church unscathed. He now thought of writing a book on ethics, and was disturbed by More's interest in the same topic, manifesting a certain jealousy of More's probable priority in publication. More was most anxious to make way for Cudworth, but the latter never published on ethics while More did (*Enchiridion Ethicum*, 1667). When finally *The True Intellectual System* reached print (vast, but still only a fragment of an even more grandiose design) its cool reception disappointed the author: the poet Dryden alleged that Cudworth had raised stronger arguments against the being of God and Providence than he was able to answer convincingly. Despite its slow start and current reputation for unreadability, *The True Intellectual System* has been reprinted several times and translated into Latin.

Making no pretence here of examining Cudworth's possible role as an historian of philosophy – for he was rather a scholar than an original thinker – the tendency of his thought seems to be sufficiently indicated by saying that it paralleled that of his friend Henry More. Like all the Cambridge Platonists, Cudworth detested and opposed contemporary manifestations of materialism. The refutation of atheistical, materialist philosophy was to be his life-work, as it was that of his friend More also.

Among the Emmanuel Platonists, John Smith in *Select Discourses* (written *c.*1650, published 1660) indicated succinctly the arguments that the Christ's Platonists were to develop at length later. The Epicureans and Lucretius had undertaken to dispose of all those mysteries in the world that give ground for superstitious beliefs; but (Smith wrote) 'they do miserably blunder themselves', through overconfidence, because explaining the world does not banish divinity from it:

For though a lawful acquaintance with all the events and phenomena that show themselves upon this mundane stage would contribute much to free men's minds from the slavery of dull superstition: yet would it also breed a sober and amiable belief of the Deity, as it did in all the Pythagoreans, Platonists, and other sects of philosophers, if we may believe themselves; and an ingenuous Knowledge hereof would be as fertile with religion as the ignorance thereof in affrighted and base minds is with superstition. (Tulloch 1874: 154–6)

Smith – recapitulating, as he says, Cicero, who in antiquity had 'stopped the wheel of this over-hasty philosophy' – contended that Epicurean atomism explained nothing since it left unexplained the principle of motion which was the origin of all phenomena. Further, how could 'these movable and rambling atoms come to place themselves so orderly for the universe, and observe that absolute harmony and decorum in all their motions, as if they kept time with the musical laws of some almighty mind that composed all their lessons, and measured out their dances up and down the universe.' To suppose that chance meetings of atoms could make a universe is in Smith's view as absurd as supposing that the learning of a book is composed in the numbers of each letter of the alphabet that it contains. 'Remove God and Providence out of the world, and then we have nothing to depend upon but chance and fortune, the humours and passions of men; and he that could then live in it had need to be as blind ... that he might not see his own misery always staring upon him' (Tulloch 1874: 157).

Smith does not seem to have attacked Thomas Hobbes directly, though this last passage reads almost as a parody of Hobbes's own account of the state of Nature, when men lived in anarchy and life was nasty, brutish and short. For the Christ's Platonists, on the other hand, Hobbes was a frequent and familiar target. One of the most recent students of the *Hunting of Leviathan* endorses Tulloch's belief that Hobbes's materialist (and, as his opponents alleged, atheistic) philosophy 'served to call forth all the energies of the [Cambridge] movement and give decision to them. While Platonism may be said to have originated the movement, Hobbism was the means of concentrating its thought and giving dogmatic direction to it. While the one was the positive the other was the negative influence which formed the [Cambridge] school' (Tulloch 1874: 25–6; Mintz 1962: 80).

The two philosophies were poles apart in character. Hobbes's philosophy was tightly argued, consistent and systematic, embracing a physical world-view, politics, psychology and ethics. It was cynical, iconoclastic, wholly unconventional, so godless that critics could not credit Hobbes's personal assertion of Christian belief (or, indeed, the evidence of his attendance at divine worship). The philosophy of the Cambridge Platonists — if a loose and nebulous doctrine may be so defined — was rambling, vague, indefinite in scope and conservative. It was well meaning, generous, wholly admirable in all its human tendencies, but also quite innocent of hard edges and clear programmes. Cambridge Platonism is one of the softest of philosophies, Hobbes's one of the hardest. Against the materialism, determinism, egotism of Hobbes, the Platonists opposed a spiritualist idealism, the assertion of free will, the qualities of goodness and abnegation of self. To Hobbes, religious faith was entirely apart from rational philosophic discussion; the Cambridge men made it the basis of their philosophy. To Hobbes, there is no such entity as 'mind' distinct from the material brain; to the Platonists, Mind was 'senior to the world, and the architect thereof'. Hobbes had rejected the great chain of being (for it is a consequence, indeed an embarrassing one, of the materialist interpretation of life that the only gradations possible between organisms are gradations of mechanical com-plexity); Cudworth emphatically reasserted that 'There is unquestionably a scale or ladder of nature, and degrees of perfection and entity one above another, as of life, sense, and cognition, above dead senseless and unthinking matter; of reason and understanding above sense, etc.' (Mintz 1962: 97). Not surprisingly, Cudworth makes much of the weaknesses of Hobbes's psychophysiology, the least convincing section of Hobbes's system since he had virtually no bricks wherewith to build. Why do we not say that the lute hears and the looking-glass sees? Are thought and imagination merely the noise produced by a mechanical system, or is it not more true — at this point the Platonists came close to the teachings of their master — that ideas are innate, indeed real and immortal? We *know* geometrical forms with the mind, Cudworth asserts, we do not simply recognize them from our experience of similar shapes in nature; 'there are no perfectly straight lines, no such triangles, circles, spheres or cubes as answer to the exactness of our conceptions in any part of the whole material universe, nor ever will be' (Mintz 1962: 99).

Like Newton later, in relation to gravitation, Cudworth had to respond to the accusation that when idealists invoke mind or God as the cause of mental or physical phenomena, they appeal to the occult. He replied, much as Newton was to do, that the phenomena themselves demonstrated the existence of mind, which is perhaps no answer. Like Henry More (but unlike Newton) Cudworth did not wish to make God immediately responsible for every event in nature that is not mechanical, partly (it would seem) because some such events seem to men absurd, monstrous or evil. The growth of the lilies of the field and the fall of the sparrow have to be supposed features of the ordinary course of organic nature, not recurrent, independent consequences of acts of the divine Will. To govern the unity of the organic world in its normal course, More imagined a universal mind, a 'Spirit of Nature' or 'Hylarchic Nature'; Cudworth's name was 'Plastick Nature'. The idea is ancient, is justified by Platonic and neo-Platonic precedents, and is by no means restricted to the Cambridge Platonists. The same idea in a different form and language is indeed Aristotelian too, and one might add Galenic, for Galen frequently personifies Nature and makes her the wise mistress of teleology. It is somewhat difficult to see how a teleological philosophy — one that detects ends and purposes in the phenomena of the universe, and believes that successive, interconnected events (such as occur in the development of an organism) are purposefully directed towards the fulfilment of these ends — can ever be reconciled with materialism and the absence of directing minds. Even the Darwinian theory of organic evolution has encountered very great difficulty over this point: how can chance, systematic order and development ever be reconciled (Raven 1953: 178–85)?

The ancient atomists, Lucretius most notably, could only by-pass this problem by making an assertion. The problem reappeared, after the long ascendency of Aristotle, with the revival of atomism in Europe during the Renaissance, being exaggerated by the eagerness of most atomistic (or, more generally, mechanical) philosophers to demonstrate that their theory of physical Nature by no means excluded God and the Christian religion. Each of the three great mechanical philosophers of the early seventeenth century — Descartes, Gassendi and Hobbes — treated the exclusion from his world-system of theistic and spiritual explanations and of occult causes in general in different ways, and so also handled differently the converse problem of reconciling his world-system with Christianity. Descartes, for example, wrote that he described the evolution of the universe *as if* it had not been created by God. As his reaction to the condemnation of Galileo at Rome in 1633 revealed, Descartes was most concerned that his own writings should not be stained by the charge of irreligion. Gassendi was himself a priest. Hobbes, most vehement of mechanists, was nevertheless invariably firm in denying the personal accusation of atheism. But none of these three mechanical philosophers permitted divine intervention as a means of reconciling chance and order.

The English empirical tradition developed with no dogmatic commitment of this kind. The groups which came together in 1660 to form the Royal Society were eclectic, Baconian, religious (indeed, devout); they were sceptical of *a priori* theories and philosophical systems; they valued highly the positive exploration of Nature by mathematical analysis, dissection, experiment and observation. Some were strongly drawn to the ideas of Descartes, others to those of Paracelsus and Van Helmont. Their bias was towards technical, detailed science rather than towards general explanations. Men of this kind – of whom Robert Boyle was the chief early exemplar, as Isaac Newton was to be the later – were content to leave some mysteries unresolved. They all opposed occult causes; they were all, more or less, mechanical philosophers; they all aspired to the establishment of closed systems of explanation for physics, chemistry and astronomy without need to invoke spirits in Nature or unaccountable powers in matter. But equally, though in varying degrees and in varying contexts, they allowed that in all the natural philosopher's investigations of Nature he must be conscious of the fact that God had created it. These philosophers, however empiricist in detail, never disregarded the wisdom of God as a relevant explanation of the overall structure and fitness of things. If philosophers asked not *how* Nature works, but rather *why* it is so well contrived to function so perfectly as it does, so that the Earth is neither too hot nor too cold, nor do comets continually crash upon it, for example, then it was legitimate for them to discern the beneficent influence of God's hand, in a way that Hobbes (at least) would have deplored. By making an appeal to divine wisdom, or what was later called natural religion, the ancient problem of materialism – the reconciliation of *chance* and *order* – was pushed aside (if not resolved) and it was possible to bring science and Christianity into complete harmony. (That the problem was far from resolution is evident from the debate between Leibniz and Samuel Clarke, the friend and partisan of Newton, about the role of God the Clockmaker.)

In this rather imprecise British way, the natural philosophers of the Royal

Society drifted away from the stark stochastic principles of ancient atomism and avoided the atheistic tendencies of Hobbism, in part impelled (no doubt) by the denunciations that emanated from the Cambridge Platonists. Henry More, as we shall see later, in due time found Cartesianism as pernicious in its philosophic implications as Epicureanism had been in antiquity: but More understood the interesting point that while the plenistic universe of Descartes must be irretrievably mechanical (since there are no spaces or gaps between the particles of ordinary matter, so that any 'spiritual substance' [in his terms or, say, field, in ours] would have to be co-terminous with matter), by contrast the universe of atoms is also the universe of space – and because God is infinite, a universe of infinite space. (Joseph Glanvill put it that 'to affirm that goodness is infinite, where what it doth and intends to do is but finite is to state contradiction.') Two consequences follow: *something* (not matter, by definition) must enable the units of matter (atoms) to act upon each other, and that *something* must be located in the space between the atoms, not in the atoms themselves. To More this something was spirit. The pleasing, necessary Platonic distinction between inert matter and active principles, between moved and mover, between the changing and the unchanging, was once more strongly reasserted.

Cudworth was not interested in this, preferring the historical method of analysis to the conceptual. He drew a distinction (probably indefensible) between the most ancient atomists (Pythagoras, Anaxagoras, Moses) who had taught the existence of a Creator and Prime Mover of all things, and the atheistical atomism developed by Epicurus. The former, he wrote, 'atomised, but they did not atheise; atheistical atomology was a thing first set on foot by Leucippos and Democritos' (Tulloch 1874: 250). Their assertion not only of the existence of atoms but of the nonexistence of anything besides atoms was a pernicious error which, Cudworth showed at length, had been rejected by all the best philosophers of antiquity including Plato and Aristotle. Belief in the existence of nonmaterial substance – that is, in an entity which can occupy space and is therefore extended, but is not matter – was therefore well established in antiquity. This entity is of course the Spirit of Nature or Hylarchic Principle. 'Corporealism', as Cudworth calls it, the doctrine of the unique existence of matter, is the true ground of atheism, one might say tautologically so.

Few have cared to expatiate upon Cudworth's notion of 'Plastick Nature' animating the organic world which seems, indeed, far from clear. Cudworth naturally insisted upon the insufficiency of corporealism to account for life in all its varied manifestations, but his attempt to depict an 'active force' operating in the inorganic world yet also distinguishing the organic from the inorganic is confused. It is an unconscious, passive agent of God: 'a dumb, patient sleepless energy, ever obedient to the divine will, and unceasingly translating it into form and action' (1874: 251). Like Aristotle, Cudworth seems to envisage the operation of a hierarchy of such 'plastick natures' – vegetative, animal, rational. Tulloch – who had no high opinion of Cudworth – speaks of his reviving 'the old Platonic dream of a soul of the world ... against the mechanical theory of Descartes' (1874: 272) yet a dream also, surely, still tinged by Aristotle's conception of an organic universe.

Isaac Newton read something of both More and Cudworth. The view that he formed of ancient atomism differed considerably from that in *The True Intellectual System*, and the extent to which he was influenced by Cudworth is, as yet, dubious. His debt to More is certainly greater. In one of those elusive drafts which are never, perhaps, to be taken *au pied de la lettre* as expressions of Newton's firm and abiding belief, he wrote that the mystic (that is, presumably, the Platonic and neo-Platonic) philosophy of the Greeks appeared to embrace atomism, and that the attachment of atomism to atheism was an absurd mistake of Epicuros.

For:

> Those Ancients who rightly understood the mystical philosophy taught that a certain infinite spirit pervades all space and contains and vivifies the universal world; and this supreme spirit was their numen, according to the Poet cited by the Apostle: In him we live and move and have our being. Hence the omnipresent God is acknowledged and by the Jews is called Place. (Westfall 1980: 511)[1]

Elsewhere, Newton seems to say that Epicuros and Lucretius were not atheists in fact, though history has made them so; their philosophy, he said, 'is true and old, but was wrongly interpreted by the ancients as atheism' – an opinion that curiously seems to place Newton close to Hobbes! An echo of Newton's rejected drafts on ancient atomism survives in Query 28 of *Opticks*, in which he refuted the Cartesian aether: 'For rejecting such a Medium, we have the Authority of those the oldest and most celebrated

Philosophers of *Greece* and *Phoenicia*, who made a *Vacuum*, and Atoms, and the Gravity of Atoms, the first Principles of their Philosophy; tacitly attributing Gravity to some other Cause than dense Matter.'[2] As is now well known, Newton also privately attributed to the most ancient philosophers – those same philosophers whom the Platonic tradition revered – knowledge of the universal law of gravitation and of its application to understanding the system of the heavens; for these reasons Newton has been placed by some scholars (perhaps too emphatically) among the Platonist followers of the *prisci theologi*.[3]

Like his friend Robert Boyle, Newton was convinced that the mechanical philosophy revived and almost universal in his own day was philosophically and theologically innocent, and on this issue he clearly stood apart from the Cambridge Platonists. It is not surprising therefore that, unlike the Christ's men, Newton does not condemn Descartes as the author of an anti-religious philosophy, firm and vigorous as are his criticisms of Descartes's physical theories. If Newton ever feared that his own system of the world might be subjected to such an attack as Henry More directed against Descartes, or such as almost all mankind directed against Hobbes and Spinoza, he left no traces of such a fear, unless one reads the General Scholium concluding the *Principia* as a pre-emptive defence against any charge of driving God out of Nature. Yet in the end he omitted from his printed texts almost all the allusions to the ancient atomists that he had drafted, even one so noncontroversial as the draft passage asserting Lucretius' awareness of the law of inertia (Hall and Hall 1962: 309–11). And perhaps few would wish that he had chosen otherwise.

Postponing from present consideration the influence of Henry More upon Isaac Newton – to be discussed later – what can be said in the most general way about the place of the Cambridge Platonists in intellectual history? That they encouraged or even founded the tolerant, inclusive movement within the Church of England known as Latitudinarianism, associated with two successive Archbishops of Canterbury, John Tillotson (1630–94) and Thomas Tenison (1636–1715), who were both Fellows of the Royal Society, is certain. A recent historian of 'rational theology' writes that the 'Cambridge men bestrode a crossroads of English religious life, and helped [to] determine the path it subsequently followed.' He sees them as 'heralds of the fresh dawn of rational theology which was presently to dominate both the horizon and the celestial cycle – at least in its English phase' (Lichtenstein 1962: Preface; 24). Modern historiography has only generalized to the whole group the degree of influence that their contemporary, Archbishop John Tillotson, assigned to Whichcote in particular: 'Every Lord's Day in the afternoon for almost twenty years together he preached in Trinity Church [in Cambridge] ... and contributed more to the forming of the students to a sober sense of religion than any man in that age' (Raven 1953: 110).

Of the ideas behind this Cambridge movement Tulloch, the first historian of rational theology, wrote in 1874 that they constituted 'the only theory of the Church which has been found consistent with Christian science, and the cultivation of intellectual fairness no less than spiritual piety and charity. Not only so, but it is the only theory not discredited by the course of civilization.' To this perhaps rather strong praise he adds: 'The noble distinction of the Cambridge Divines is that they at once rationalised religion and vindicated its distinctive reality' (1871 c: 465–6; 470). And he opines that it is impossible to overrate their services. The faults of the Cambridge Platonists, as Tulloch saw them, were incidental to this overwhelming merit: their disregard of method, their lack of historical criticism, their fanciful and credulous arguments. As Tulloch states them, such weaknesses seem rather to belong to the Christ's men than to their Emmanuel predecessors. The most recent tiller of this well-dug ground, Margaret Jacob, endorses Tulloch's assessment of the importance of Cambridge Platonism to the Latitudinarian movement, and of its role as a mediating force between pure theology and abstract science. The Cambridge Platonists, she writes, helped to create the 'social ideology of the liberal Anglican establishment' which was supported by the Newtonian philosophy of nature, and so helped further to give English society and intellectual life their distinctive flavour during the early eighteenth century (1976: 271).

Therefore there is little reason to doubt that historians have given a national, if rather short-lived, significance to the Cambridge school. The not infrequent reprintings of their writings teach the same lesson. But what of a wider and perhaps more permanent reshaping of the course of philosophy? This seems harder to establish: the Cambridge Platonists appear merely to fill conveniently a chronological interval between Hobbes and Locke in the English tradition, though in perverse and ineffectual contradistinction from both. (Yet one should remember that Locke records in his journal a debt to Cudworth's *Intellectual System*: see Raven 1953: 110; Ayers

1981, esp. 233–51.) One version of the history of English philosophy might see the Cambridge Platonists as critics of Hobbes who were themselves severely taken to task by Locke; or as simply provoking the good-humoured refutation of Leibniz who, nevertheless, had some esteem for Cudworth and appreciated the poet in More. Cassirer found it difficult to express the historical standing of Cambridge Platonism sharply. It was, he wrote, 'not a mere literary curio, but an integrating factor, an important and necessary stage in the growth of the modern mind' (1953: 7). The reader may interpret 'stage' as 'passing phase': as marking a necessary antithesis to the materialism of Descartes and Hobbes (whether real or apparent) before philosophy moved on to a more assured and permanent stance. But Cassirer had already written: 'Neither in intellectual scope nor in immediate influence is this work [of the Cambridge school] comparable to the great spiritual forces which formed the modern world picture' (1953: 1). This may well be true, but the comparison is too open-ended. If the wielders of the 'great spiritual forces' were, for example, Galileo, Descartes, Leibniz, Locke and Newton, then one may agree that each of these was a man of greater intellectual stature than any of the Cambridge Platonists. But does it make any sense to compare Culverwel with Leibniz, or Cudworth with Galileo? Their problems are too unlike. Any such comparison would seem to reduce the Cambridge men to the level of mere uncomprehending reactionaries, at least in the fields of philosophy and science. Yet this was not wholly Cassirer's view of them, since he also wrote: 'Cambridge Platonism represents a coherent philosophical position ... which continually recurs as a central theme ... the name "Cambridge School" refers to a certain line of thought of independent force and significance which is deliberately opposed to the prevailing direction of English thought in the seventeenth century' (1953: 6). Opponents of 'progress', then, but not uncomprehending or insignificant exponents of a minor tradition that 'constantly recurs'. One has to suppose that Cassirer meant by this the tradition of idealist and spiritualist philosophy.

It would be out of place to pursue this theme here. It is time to look more closely at Henry More himself. But in the last reckoning one cannot conceal the fact that so far as natural philosophy (and *a fortiori* natural science in the modern sense) are concerned, the Cambridge Platonists were at the end of the road. Leibniz was surely right on this score. Admiring Plato rather as a poet than as a philosopher, critical of the neo-Platonists for 'scurrying after the miraculous and the mystical', he found fault with More for his neglect of efficient causes and his too easy recourse to miracles whenever explanations in terms of mechanical principles were not very obviously to be preferred. The mechanists, according to Leibniz, were in error by their neglect of metaphysics and scorn for all explanations outside the reach of sense-perception, but world-souls and ideas independent of things were mere poetic fictions (Cassirer 1953: 153–5). If Leibniz was right, so were Boyle and the other English philosophers of the experimental-mechanistic persuasion whose accounts of phenomena More rejected because they contravened his own spiritual-mechanistic principles. We may read the Cambridge Platonists as furnishing historians with strong indications of the difficulties aroused in the minds of intelligent, thoughtful and well-meaning men by the application of the mathematical and mechanical philosophy to the development of experimental science; we may read them also as providing subtle, clever and learned expositions of very ancient cosmological principles which were not without relevance to the intellectual life of the age of Newton, serving still (as they always had) to express humanity's admiration of the order and mystery of the universe. We may read them also as furnishing historians with a salutary reminder of the true complexity of intellectual history. But we cannot read them as the precursors of some great philosophical movement whose maturity was yet to come.

5

Henry More, Man of Paradox

ROM SIR THOMAS MORE , the former Lord Chancellor of
England executed in the Tower in 1535, descended two
grandsons named Henry, who were cousins and Catholics. Neither of these
was the Cambridge don, born in 1614, who came of a family of 'estate
and fortune' long established in Grantham, Lincolnshire. A memorial in
Grantham Church to Gabriel More Esq. (*c.*1635–98) records that he was
the nephew of Henry More DD of Christ's College, Cambridge, who was
'by his learned writings in divinity and philosophy one of the greatest
glories of our church and nation, and who tho' eminently charitable in his
life time, at his death left an honorable addition of estate to his said Nephew.
He was the last branch of this worthy stock' (Turnor 1806: 25).[1]

Isaac Newton, a generation younger than Henry More, born at Wools-
thorpe, only an easy ride from Grantham, like More began his education
in the Free (or Grammar) School at Grantham. During seven years Newton
lodged with a certain Mr Clark next to the George Inn at Grantham: this
man was an apothecary, whose brother, a teacher in the Free School, had
been a pupil of More's at Cambridge. With one brother or the other More
lodged when he paid a visit to his native town. Inevitably, therefore, More
was known to the youthful Newton indirectly, probably directly, long
before the boy became an undergraduate in 1661 (Turnor 1806: 176;
Nicolson 1930: 98).

Little is known of More's early life, apart from his spiritual development as narrated by himself in the Preface to his collected works in Latin. Up to his fourteenth year of age he was brought up by 'Parents and a Master that were great Calvinists (but withal very pious and good ones).' He always had serious thoughts about religion: 'even in my earliest Childhood, an inward Sense of the Divine Presence was so strong upon my Mind, that I did then believe, there could be no Deed, Word or Thought hidden from him' (Ward 1710: 5; 7).[2]

At 14, he was sent from Grantham to Eton College, where he underwent in the form of a religious conversion a total revulsion from the doctrine of predestination, believing that if he should conduct his life as a good man, and as though one of the Elect, God would not possibly reject him. At this time, it is said that an uncle placed *in loco parentis* over the boy threatened to flog out of him his 'immature forwardness concerning the mysteries of necessity and free-will' (1710: 6), but More never relaxed his grip on the principle of religious voluntarism. In 1631 he was admitted to Christ's College, Cambridge, a College inclined to Puritanism, where his 'most successful and vigilant' tutor was William Chappell, who was also Milton's tutor (and, according to report, flogged his pupil). More was also advised by the greatest Christ's man of the age, Joseph Mede. When Chappell questioned his pupil about his assiduity in study, More insisted repeatedly: 'I desire so earnestly to know, that I may know.' He plunged

> over Head and Ears in the Study of Philosophy; promising a most wonderful Happiness to myself in it. Aristotle, therefore, Cardan, Julius Scaliger and other Philosophers of the greatest Note I very diligently perused. In which the Truth is, that I met here and there with some things wittily and acutely and sometimes also solidly spoken: Yet the most seemed to me either false or uncertain, or else so obvious and trivial, that I look upon myself as having plainly lost my time in the reading of such authors. And to speak all in a word, those almost whole Four Years which I spent on studies of this kind ... ended in nothing, in a manner, but mere Scepticism. (1710: 10)

From the more orthodox philosophers, in a state of doubt not unlike that recorded by Descartes on leaving La Flèche, More turned in his longing for a more divine and metaphysical philosophy to the 'Platonick Writers, Marsilius Ficinus, Plotinus himself, Mercurius Trismegistus; and the Mystical

Divines' teaching that the Divine Inflation could only follow purification of the Soul; 'none so pierced and affected me,' More continues, 'as that Golden little Book, with which Luther is said to have been wonderfully taken, viz *Theologica Germanica*.'[3] After passing through another spiritual crisis, More was so fortunate as to attain a 'most Joyous and Lucid state of Mind; and such plainly as is ineffable'. It does not seem that More was ever seriously troubled by spiritual problems again during the remainder of his life (1710: 12).

It was presumably at this period in his Cambridge studies that More became convinced of the excellence of Plato's philosophical teaching; 'I do with my pen, my mouth and from my heart praise that excellent philosophy of Plato's as the most consistent and coherent Metaphysical Hypothesis, that has yet been found out by the wit of Man.' The somewhat romantic tone of this declaration fits with More's fervent yet unanalytical admiration of the Master: 'It is rather because of More's reverence for Plato than from any clear understanding or following of Platonic doctrine that he is given the name of Platonist' (MacKinnon 1925: xxvi). Inevitably, as Plato was assimilated to the English Broad Church, the precision and individuality of his conceptions were blunted: in More's hand the doctrine of Ideas becomes a theory of the omnipresence of Spirit. In fact, to adapt a familiar *mot*, one may say that for a philosopher More was an excellent poet, and that for a poet he was a fine philosopher.

Henry More was elected a Fellow of Christ's in 1639, the year following the death of Mede, whose natural successor he was to be. The historian of the College, John Peile, notes the curious similarities between the two men: both were (to some extent) philosopher, mathematician, theologian and interpreter of the Book of the Revelations (1900: 32). More called his senior 'that incomparable interpreter of prophecies', a judgement endorsed by Isaac Newton who referred to him as the 'judiciously learned and con-scientious Mr Mede whom . . . I have for the most part followed'. In addition, both Mede and More were successful and influential College teachers and, like More, Mede had been a rationalist and a Latitudinarian: 'I cannot believe that truth can be prejudiced by the discovery of truth,' he declared. Both in the true Protestant tradition were assured that 'Popedom is the Beast and Rome the Whore of Babylon.' Both men refused high advancement within the Church, choosing rather to live and die within College walls.

We possess few details about the first years of More's tranquil academic life. He desired little but books, and was possessed of an unambitious calmness of spirit. His mind was, in the words of his only biographer, enriched by the unusual 'Influences or more than ordinary Illapses of the Holy Spirit; the uncommon Notices, or secret Intercourses, of some of the good Genii or Spirits from above, which one may perceive he sometimes enjoyed' (Ward 1710: Preface). Again, Ward says: 'From the Beginning of his Time, all Things in a Manner came flowing to him, and as the Beames of the sun in the cool early Day, rose and shone upon him with their Golden and unexpressible Light.' This happy complacency is borne out by More's own affirmation that his sense of joy in God's love and his creation was so high that he felt himself *incola coeli in terra* (a celestial being dwelling upon Earth) (1710: 23, 54). He felt the seal of God upon him, who had made him

> full Lord of the Four Elements; and hath constituted me Emperor of the World. I am in the Fire of Choler, and am not burn'd; in the Winter of Phlegm and am not drown'd; in the Airy Sanguine, and yet not blown away with every blast of transient Pleasure, or vain Doctrine of Men. I sport with the Beasts of the Earth; the Lion licks my hand like a Spaniel, and the Serpent sleeps upon my Lap and stings me not. I play with the Fowls of Heaven; and all the Birds of the Air sit singing on my Fist. All these things are true in a sober sense. (1710: 45)

One may set beside the self-conscious echo of the Authorised Version in these prose lines the same sentiment as given by More in classical dress in the *Philosophical Poems*, equally ecstatic:

> When I myself from mine own self do quit
> And each thing else; then an all-spreaden love
> To the vast Universe my soul doth sit,
> Makes me half equall to All-seeing Jove.
> My mightie wings high stretch'd then clapping light
> I brush the starres and make them shine more bright.
> (Tulloch 1874: 316)[4]

Elsewhere too More returns to somewhat the same feeling rather less fancifully when he remarked of the wonders of Creation: 'A good man could be ready in his own private Reflections, to kiss the very Stones of the Street' (Ward 1710: 56).

A tendency to excessive warmth of emotion and extravagance of language was perceived by More in himself. He possessed, he confessed, a 'Natural touche of Enthusiasme in his Complexion; but such as (he thanks God) was ever governable enough, and which he had found at length perfectly Subduable. So that no Person better understood the extent of Phansy, and Nature of Enthusiasme, than he himself did' (1710: 45).

In 1642 Henry was presented by his father, Alexander More, to the living of Ingoldsby in his native county (1710: 56). After John Worthington's expulsion from the Mastership of Jesus College, More gave him this incumbency; later he presented the advowson of Ingoldsby to Christ's College (Crossley 1855: 218–23).[5] In 1654 (according to tradition) More refused to stand candidate for the Mastership of his College, preferring to pass 'otherwise his time within those private Walls; it may be as great a Contemplator, Philosopher and Divine, as ever did, or will in hast visit them' (Ward 1710: 61). It seems that he was never officially Tutor of the College though he took a few pupils and was meticulous about the rules of residence. More's refusal of office came from an honest assessment of himself: 'I have measured myself from the Height to the Depth; and what I can do, and what I ought to do, and I do it' (1710: 78). He was no man of affairs and public life; Ward says his voice was weak. Ward also tells us – presumably from personal knowledge – that More 'was once for Ten days together, nowhere (as he term'd it) or in one continual fit of contemplation: During which though he eat, drank, slept, and went into Hall, and convers'd yet the Thread of it for all that Space of Time was never once, as it were, broken or interrupted' (1710:42). Isaac Newton, as is well known, experienced similar fits of deep abstraction, during which he did not know whether he was making his way to Hall or Chapel. More was sufficient of a mystic – though Lichtenstein has denied him this title in the strict sense of the word – to believe that the world of the inner vision is more real and permanent than the world in which men drink, sleep and dine in Hall: 'The whole Life of Man upon Earth, Day and Night, is but a Slumber and Dream, in comparison with that awakening of the Soul that happens in her Recovery of her Aetherial or Celestial Body' (1710: 71).

No man's character and conduct of his life has been more admired in his own time. William Outram, also a Fellow of Christ's, called More 'the Holiest Person upon the face of the Earth'; another said 'He was an Angel rather than a Man; and would be amongst the Angels of God for ever'

PLATE 2 Christ's College, Cambridge, from David Loggan's *Cantabrigia Illustrata*, 1690, Plate No. 25, 'Collegium Christi'. (Reproduced by kind permission of The Bodleian Library, Oxford: ref. Douce L.Subt.26)

(1710: 79). He himself only described his manner of living as 'Harmless and Childish'.

More's Christian faith was simple and basic. The avoidance of intellectualism and doctrinal pettifogging typical of Cambridge Platonism is well expressed in the Preface to his *Mystery of Godliness* (1660):

> My onely solicitude ... was to corroborate that Faith that is plainly propounded to us by Scripture, which is sufficient to Salvation, and to exalt that Life that has lyen dead and buried for these so many Ages under a vast heap of humane Inventions, useless and cumbersome Ceremonies, and unpeacable Opinions.

Again and again More insists on the simple doctrine that the harmony of the soul is gained by withdrawal from the world and abnegation of self; so to John Worthington in 1668:

> Not to seek to please a man's phansy in outward contents, is the next way to live in perpetual peace and contentment, and in enjoying God and a good conscience, to enjoy all things. Whosoever will be happy either here or hereafter, the most certain and compendious way is, to rend himself from all the magick of this outward life, and industriously to decline all such attractions, and to be as ready to thwart a man's self, as providence is to thwart one, which assuredly, as often as she does, is at least to the good, a design of her to drawe the soul to that inward power and life, wherein consists our main happiness both in this world, and that which is to come. Neither many words, nor much knowledge, nor the voluminousness of books, which are the disadvantages of our academical education, are any thing to this; but it is the perpetual taking up of the Cross, and constant endeavour to shun a man's own will and appetite, that leads directly to this resurrection of life and peace and joy; and is the health both of soule and body. (Christie 1886: 279, 9/9/1668)

Like many other Fellows in both Universities, More survived the vicissitudes of the Civil War, Commonwealth, Protectorate and Restoration. He was fundamentally apolitical, as he was broad in his religion; King, Republic or Protector was acceptable to him, so long as the College was undisturbed. With Thomas Bainbridge (twelfth Master), William Moore and Ralph Widdrington he was left unscathed by the Puritan purge of the University in 1644, by which Joseph Beaumont, later to be Master of Peterhouse and an enemy of More, was expelled from Christ's. The numbers of annual admissions of students declined during the war, but recovered to between fifty and sixty when peace was restored and life went on with its former round. More's chief friends among the Fellows were William Outram, later Archdeacon of Leicester, George Rust (p. 61) and Thomas Standish (c.1633–1714), longer a Fellow of Christ's than anyone before or since, by whom many of More's books were to be presented to the College Library (Peile 1900: 164–5, 172).[6]

Ralph Cudworth (p. 70) came as Master to Christ's from Clare Hall, where he had been intruded Master in 1645 when only 28 years old. It has been plausibly conjectured that he had been unable to secure from the College the normal revenues of his office, since he pleaded poverty as the reason for seeking this translation to Christ's.

The course of the relationship between More and the new Master is not clear, but the former had dedicated a book to Cudworth in the year before his election, which he may be supposed to have favoured.[7] More's association with Cudworth, theologically, philosophically and in the business of the College, involved him in the attacks made upon the Master by Ralph Widdrington, long Tutor and close contemporary of More's in the College. Widdrington also long served the University as Public Orator, and became Professor first of Greek (1654) and then of Divinity (1673). The origin of Widdrington's animosity towards the Christ's Platonists seems to have been forgotten: whether he resented Cudworth's election, or the difference was one of party only. Samuel Pepys's brother was an undergraduate at Christ's at this time; Pepys recorded in his *Diary* a report that Widdrington 'did oppose all the Fellows in the College, and that there was a great distance between him and the rest', which dissensions he feared might be to his brother's disadvantage. Widdrington had in fact just failed (1660) to eject Cudworth from the Mastership, which was again confirmed upon him after the Restoration; Cudworth in turn failed to remove Widdrington from his Fellowship during 1661–2, after the latter's appeal to the Privy Council. This was far from being the end of the matter and More continued to relate to his friend Anne, Lady Conway, the troubles that beset him as Cudworth's friend and ally, which he feared might lead to the loss of his own Fellowship. In 1664 More's adversaries 'had like to have given me a long play day. They articled very peevishly against Dr

Cudworth, butt their chief design so farr as I see was against me. This is no more than [a] stratagem of the old Serpent, to putt Christians into beasts skins and then to worry them' (Nicolson 1930: 235).[8]

The formal ground of complaint against More was of excessive absence from the College, beyond the statutory allowance, and More admitted an excess of two or three days in the year. He recognized, however, that there was really a 'plott against Dr Cudworth himself', if not against all his friends, springing from the 'aking tooth' of their opponents. More found himself

> rayl'd at and bluster'd against for an Heretick, and doubt not but that there are [those] that eagerly desire I were so, though they *feigne* so great a Zeal against Heresy. But for my own part, I thank God, I finde the clearness of my owne wayes and the integrity of my spiritt above the pedantry and calumnys of any man whatsoever. (Nicolson 1930: 220, 31/12/1663; 238, 238,5/6/1665; 242, 29/6/1665)

More predicted that Cudworth would suffer no harm from Widdrington's malice, and indeed nothing resulted from his complaint to the Archbishop of Canterbury that Christ's had become a 'seminary of Heretics'. In July 1665 More was still complaining to Lady Conway that 'They push hard at the Latitude Men as they call them, some in their pulpitts call them sons of Belial, others make the Devil a Latitudinarian, which things are as pleasing to me as the raillery of a jack-pudding at one end of a dancing-rope' (1930:243, 10/7/1665).[9] But still nothing happened. More and Widdrington were compelled to live together as Fellows of Christ's till their deaths in successive years, long after.

At about this time, as already mentioned, there was a short breach in the harmony between More and Cudworth because the Master, intending to publish a work on ethics, heard that More was about to send to the press a book on the same topic. If he, Cudworth wrote, 'should violate friendship in this kind, it would more afflict me than all that Dr Widdrington ever did, and make me sick of Christ's College and of all things in this life, (Peile 1900: 181; Crossley 1885: 157ff., January 1665). More was less concerned (he had other works in hand, such as his *Divine Dialogues*, 1668); he was quite ready to let Cudworth publish first. But the Master's work was far from ready: indeed, his book was never to be printed at all, though a large fragment of it still exists in manuscript. Accordingly, More's *Enchiridion*

Ethicum – one of his least interesting books – appeared in 1668.

More had begun his career in print in 1642 with *Psychodia Platonica: or a Platonicall Song of the Soul, consisting of foure severall poems*; this had been composed two years previously and, no doubt, reflects the first flush of his enthusiasm in his discovery of Plato. This work was reprinted, enlarged, with others in *Philosophical Poems* (1647), dedicated to his father. As a metaphysical poet, More sang of the 'Infinity of Worlds, the Praeexistency of the Soul,[10] the Immensity, nay Infinity both of Time and Space, the Soul's Immortality, the adorning of the Earth with Universal Righteousness in due time, (Ward 1710: 24). More's prose writings began in 1650 with an attack (followed by a rejoinder to the reply, bearing the title *The Man-Mouse taken in a Trap*) on Thomas Vaughan (1622–66), the mystic alchemist. More's first major book, *An Antidote against Atheism*, appeared in 1653 and perhaps his best-known of all, *The Immortality of the Soul*, in 1659; meanwhile More's fluent pen had also produced *Conjectura Cabbalistica* (p. 110) and *Enthusiasmus Triumphatus*, 1656. The *Poems* were admired enough to merit immediate reprinting. More's book on ethics was also quickly reprinted.

Thus the name of Henry More was already becoming known when he first met Viscount Conway in London in 1653, to whom he dedicated *The Immortality of the Soul*, noting that the idea of writing upon this subject had first come to him while reading Descartes's *Passions of the Soul* with Lord Conway in the Luxemburg Gardens in Paris 'to pass away the time'. That must have been in 1656. Far more vivid and creative was More's friendship with Anne Finch, who was to become Lady Conway. One of her elder brothers was Heneage Finch, the statesman who became first Earl of Nottingham in 1681; another was John Finch, who entered Christ's College in 1645 and became More's pupil. After a migration to Oxford John Finch proceeded MA from Christ's in 1649, later taking the MD at Padua, where he served for a time as English Consul. Then he was appointed to a Professorship at Pisa by the Grand Duke of Tuscany. Through him More met his sister Anne, presumably in London, and at once accepted the role of mentor to this learned young lady, who had already studied Plato, the Platonists, and the cabala. With More she went through Galileo and Descartes. She married Lord Conway in 1651; the earliest surviving letters in her correspondence with Henry More are from the previous year. They met sometimes in the Conways' home at Kensington House (later improved into Kensington Palace), but More made longer visits to the Conways' country house at Ragley in Warwickshire; at first, as he told Lady Conway, More 'had rather wayt upon you in London then in Warwickshire, because I have naturally an averseness from being in strange places and conversing with such as I am not acquainted with, especially great persons'. With greater familiarity, and with fewer 'great persons' resorting to Ragley, which became Lady Conway's private retreat, More developed a great attachment to the countryside, the house and its chatelaine. Ragley Hall became More's 'true home, to which he hastened on the first day of his release from academic duties' (Nicolson 1930: 110, 9/6/1655; 112).[11]

The house is in the southwestern part of the county, close to Alcester. It is a lightly rolling pastoral country, lacking any special charm. Both the Conways were to be buried in the nearby parish church of Arrow, together with their infant son, Heneage, who died in 1660; only the child's burial is now recorded by a small brass plate. The church has of course been greatly restored and enlarged since the days when Henry More must have walked from the Hall to worship there. The Hall too would be unrecognizable to him. After Anne Conway's death in 1679, upon the Earl's remarriage, the house was totally reconstructed for him by Robert Hooke, Surveyor of the City of London after the Fire. His house in turn was improved by the Seymours, Earls and Marquesses of Hertford, who succeeded the Conways there. In the eighteenth century also the park was redesigned by 'Capability' Brown, leaving nothing in the immediate prospect that was familiar to the eyes of Henry More and Francis Mercury van Helmont.

How fondly More regarded both brother and sister appears in the letter (2 November 1651) wherein he assures Anne Finch that their joint desire for a continued correspondence with him can be no greater than his own:

> For that candour, freeness and perspicacity of witt which I have observed in your brother is so eminent in yourself, that whyle I converse with you, the better part of him is still present here in England, and will salute me at every arrivall of your letters to my handes. And therefore you may easily speed when, seeming to beg a favour, you really confer one of the greatest value upon me ... I hold myself bound as well in point of religion as civility, to afford you what service lyes in my power. (Nicolson 1930: 53–4)

His willingness to act as a mentor to Finch's sister was founded on no 'confidence in my abilityes in any grand points of Religion, as if my judgement were consyderable', but rather it will be for him as a friend to 'discuss with you what you shall be pleased to propound, but yourself must chuse according to the present Light of your own minde'. Then he goes on to state the Platonic doctrine of the true philosopher as a 'holy temple': 'the thirst after knowledge is ever dangerous till the divine life has its birth in a man, and so the soul becoming divine, God in man as I may so speake examines his works over againe ... And then all the inquisitions after knowledge are as safe as sweet' (1930: 53–4).

This lady, to whom More wrote so many long letters of philosophy and religion, with many glances at Cambridge affairs, public matters and ghostly visitations, became the brightest light of his life. His feeling for her became, in the words of the editor of their correspondence, a decorous love 'as reverent as Dante's for Beatrice'. One of the warmest expressions of More's emotions is in a letter written immediately after his return from Kensington in early January 1656, overflowing with

> the pleasure of your Ladiships company from which I reape so great satis-
> faction, that it makes that Life, which used to give me the greatest content,
> very dead and heartless to me. When I first came into my chamber, methought
> it look'd very sad and desolate, and I found it no complement, that after my
> converse with so noble a friend, my retirement to Cambridge would be like
> coming into an obscure cottage. After a long walk in the sunshine and snow
> all things afterwards looke darke. (Nicolson 1930: 128, 7/1/1656)

Never had he so tasted the sweetness of friendship before, nor can he command himself from most affectionately loving her, whom it is his duty to honour and adore. And love itself is far from being incompatible with the highest objects. 'But discretion bids me temper myself and absteine from venturing too farr into so delicious a theme' (1930: 128). When he hoped to welcome Lady Conway to Cambridge in 1657 he assured her 'it will be the joyfullest day that ever I did see or am ever to see in Christ's College' (1930: 143). Lady Conway's esteem for More was hardly less than his for her: projecting her return from a long stay in Ireland with her husband in 1664 she assures More 'there is no body there [in England] I shall have as great passion to see as yourself ... I cannot forbear pleasing myselfe sometimes with the thoughts and discourse of what contentment I should take in your company' (1930: 224).

Though to a modern reader John Finch seems vain and pompous, his lifelong friend (and fellow pupil under More)' Thomas Baines, dull and pedantic, and the lifelong marriage of these two minds questionable, for Henry More his friendship and social intercourse with the Finch family was throughout the middle years of his life an especial source of delight and inspiration. As he told Lady Conway in September 1660,

> I can assure you my minde has been hitherto wholely with your Ladiship and the rest of that learned Conclave, I mean my Lord [Conway] and the Italian Doctours [Finch and Baines], and phansy myself present at all your excellent Discourses so strongly that I can scarce find myself here at all ... if I know myself aright, I am nothing else but an *Aggregate* of my friends so that they that are the best and choicest of them are the greatest part of myself: and that you are the chiefest of all I must with thankfulnesse acknowledge, nor can anyone deny it.

Accordingly, he goes on, nothing but force would keep him from Ragley and the best company he knows. 'I pray God', he concludes, 'give good success to your new course of Physick that it may at least procure some ease, if not a perfect cure' (Nicolson 1930: 164–5).[12]

From at least the time of her acquaintance with More, and probably for some years before, Lady Conway had suffered from the most agonising and persistent headaches, or attacks of migraine. These episodes, during which she was completely incapacitated, became increasingly frequent as the years went by, to the point where More was able actually to converse with his lady perhaps only two or three times during a long visit. In the 1660s she completely abandoned her London life, and as her husband was deeply immersed in public affairs in Ireland and Whitehall there was a virtual separation of husband and wife. Medical treatment completely failed to give relief. In May 1656 More went with Lady Conway to France, where her husband joined them in June, with the object of her submitting to the operation of trephination; in this very ancient surgical process a small circle of bone is cut out from the skull in order to relieve the presumed pressure on the brain. Because the Parisian surgeons declined to operate, the journey was in vain – unless the writing of *The Immortality of the Soul* be reckoned a consequence of it. While in France, More also formed the project of publishing an edition of the letters of Descartes, but in this he was forestalled by Claude Clerselier (1657). The party returned to England in September.

The second attempt to cure or alleviate Lady Conway's distressing ailment by bringing over from Ireland the 'stroker', Valentine Greatrakes, only marginally involved More. In a procedure reminiscent rather of primitive medical practice than of modern physiotherapy Greatrakes massaged out of the bodies of sufferers the evil principle that caused their pain; there were many testimonies to the success of his ministrations, for which he claimed no magical or miraculous powers. His visit to Ragley in January 1666 brought no relief to Lady Conway, as Greatrakes frankly admitted. On the other hand, Ralph Cudworth's little son was said to have received 'reall good' from his hands.

More's obligations to his College, which limited his visits to Kensington and Ragley, prevented his accepting an invitation to accompany the Conways to Ireland in 1661, when Lord Conway was given an official post there (1930: 183, 31/3/1661).[13] Similarly on another occasion ten years later More had to delay his departure for Warwickshire: 'For the King's coming to New-Markett and the report of his intention to come to Cambridge after the Duke of York with the Dutchesse of Cleaveland had been here made me stay one week longer to kiss the King and Queen's hand as well as the Duke and Dutchesses.' The royal couple did not make the visit but the delay enabled More to give a dinner at Christ's for Francis Mercury van Helmont, who was to become an important figure in Lady Conway's circle (1930: 322, 13/10/1670).[14] Because of his influence in Ireland, Lord Conway was able tentatively to offer to either More or Cudworth an important preferment in the Irish Church: 'We have some Bishopricks in the hands of very old men almost 100 yeares old, which are worth £1500 a year' (1930: 298, Lord Conway to More, 9/11/1669). More declined all such offers of church office, including the Provostship of Trinity College, Dublin, and two bishoprics, until in 1676 he agreed to become Prebendary of Gloucester; this post, however, he at once resigned in favour of his friend Edward Fowler.

More's letters to Lady Conway are often coloured by his 'long and anxious study and sedentary life' in College. Particularly during the 1670s, when he was engaged in translating his own writings from English into Latin for the *Opera omnia* that appeared in 1679, he complains of his self-imposed labours of study: thus, in 1663 'I have been in such a perpetuall career of studying and writing, ever since your Ladiship left England, that I have consumed the strength of my body, and wasted and wounded my health very sore' (1930: 220, 31/12/1663). At an earlier time, when there was smallpox and other sickness prevalent among the Fellows of the College, More continued his studies with even greater resolution: 'I think that the heating my spirits with study kept of[f] the disease' (1930: 195, 16/11/1661). And he once told Robert Boyle that he had cured an attack of the quartan ague 'by this diatrion, fasting, sack, and mathematics, especially Oughtred's *Clavis*' (1930: 265, 27/11/1665). (This was written in November 1665, soon after the onslaught of the Great Plague.) If More mastered William Oughtred's *Clavis Mathematicae* he clearly knew something of algebra as well as geometry.

Henry More's biographer, Richard Ward, avers that he gave 'Excellent Lectures' to his pupils and introduced them to 'Piety and Instruction from the Chapter that was read on Nights in his Chamber' (1710: 192). We may well believe that at a time when a paralysis of negligence was already beginning to afflict the Universities, More retained the conscience of earlier times. But by 1675 he was not eager to take pupils, for he told Lady Conway this and explained that he had only accepted Sir Robert Southwell's nephew as a pupil because the boy had been sent to Christ's College for More's sake, and his father was a neighbour (Nicolson 1930: 400). The sizar whose duty it was to assist More as his amanuensis was often a cause of irritation: 'I have been hugely bejaded with my former and this sizour more than any hackney, which is a vexation not [to] be understood by any but those that are putt to it' (1930: 333, 29/5/1671). And again the same simile only a little later: 'to have such a sizour to write, and a tired Cambridge hackney to ride on when a man is belated on his journey are calamityes much alike' (1930: 352, 5/2/1672). Evidently More – now aging indeed – was no great horseman, since he warned the people at Ragley, where he was expected, that 'unless my Lord has Sorrell still or a horse as sober and governable, I had as good adventure with a bad guide as an unruly horse' (1930: 324, 13/10/1670). The figure was one of which More was fond: in the early days of his correspondence with Lady Conway he warned her against metaphysics from his own experience: 'never man, I think, having been more pittifully rid[den], and hackneyed with the witchery of speculation than myself' (1930: 76, 4/4/1663).

Something of the diversions of College life also appear in the *Conway Letters*. Ward tells us that More was regular in his attendance at Hall and Chapel, but on Fridays dined in private finding that fish did not suit his stomach. He drank ale and wine in moderation and enjoyed a good dinner with amusing talk; Ward called him one of the 'Merriest Greeks' in the College (1710: 119). In September 1670 he told Lady Conway that he had taken up his old lute from the bed's head, where it had hung untouched for twenty years, and had had it refurbished so that 'now I do play with no small pleasure to my self some half a dozen my best and hardest lessons I have in my book. And within a [short] time I think I shall have them all better than ever I had them,... a proper solace in this drudgery I labour under' (Nicolson 1930: 307, 15/9/1670). He kept up his music for a time at least, and enjoyed other proper diversions:

> I play at bowles ordinary twice a day, and after a rubber after supper I take a walk into the fields, and at my returne play some lessons on my lute, and after fitt myself for bed. This and translating [for the Latin *Opera*] is now the whole course of my life, but when I am interrupted with companie. (1930: 342, 14/7/1671)

At this time each new Fellow was required to present a pair of bowls to the College upon his admission. Evening walking was evidently a habit with More, as he had more than a decade earlier remarked on his custom of taking 'a serious walk of two houres long in the College Orchard', relics of which (like the bowls) still existed in my own day (1930: 181, 4/12/1660).[15]

The daughter of More's old friend, the Master, reminds us that pedantry and dullness are inseparable from academic life: 'Methinks,' she wrote to her friend John Locke, whose 'governess' she was, 'that it is intollerable to live in a Place where it is not possible to spend one Houre in an agreable Conversation, without being beholden to the Dead or the Absent Liveing for it, and that too in one of the most famous Universities in the World' (Locke 1976, II: 518, 23/5/1682).[16] Damaris Cudworth, a young woman who had lived long with the Platonists of Christ's, had only recently returned from a stay in London, a city which she (like More at times) found a good deal more amusing than Cambridge with 'the Impertinent Wranglings of its Schools', though even there one might occasionally hear something edifying or at any rate diverting (1976, II: 736, 15/9/1685).[17]

In an unhealthy age, More was seldom distracted by illness from his constant literary labours, and lived to a ripe maturity. Ward writes that More thought himself to possess a body 'built for an Hundred Years, if he did not over debilitate it with his Studies' (Ward 1710: 41). He habitually enjoyed the blessing of sound sleep 'so that the Falling of a House could scarce wake him'. Ward describes him as tall, thin, pale, with a hazel eye as 'Vivid as an Eagle'. He had a strange conviction that his body carried a natural scent of flowers, so that towards the end of the winter or beginning of spring even his discarded clothes bore a 'flowery and aromatick odour'. More believed that 'Mirth and Chearfulness are but the due Reward of Innocency of Life,' which he maintained by his isolation 'Archimedes-like ... busie in his Chamber, with his pen and lines, so as not to mind the Bustles and Affairs of the World that were without'. Evidently Ward carried a mental picture of him working at geometry (Ward 1710: 143, 191, 211, 214).

More relates that *The Grand Mystery of Godliness* (1660) was written because of an illness which he thought would finish him, when he solemnly swore to amend his pen if he was spared (Gabbey 1982: 223–4). This was perhaps in the autumn of 1653, when he was 'troubled with a tough phlegm in my stomach and head, with the scurvy, the spleen, sinking of spirits, weakness of my legs, wasting of my flesh, and heaviness in my head, and perpetuall sleepiness, so that I suspect myself not far off from an Apoplexy, which is an easy death' (Nicolson 1930: 82, 2/1/1654).[18] Again, in May 1659, he suffered from a distemper that 'began so roughly with me that the Physicion knew not whether I would live or die.' In the fashion of his day More kept his friend Lady Conway fully informed not only of these more serious episodes of ill-health but his routine self-medication by vomit and purge (Nicolson 1930: 91, 158).

The association with Lady Conway, whose leanings towards mysticism and religious enthusiasm were stronger than his own, probably stimulated More's preoccupation with the immaterial and the abstruse, the occult and the supernatural. So did other friends, notably Joseph Glanvill. Through Lady Conway, More became deeply involved with such mystic figures as Francis Mercury van Helmont, Ezekiel Foxcroft and the Baron Knorr von Rosenroth. Van Helmont passed the chief part of his time during his long sojourn in England with Lady Conway at Ragley. The learned Mrs Foxcroft, mother of Ezekiel, was Lady Conway's chief *confidante* and amanuensis. Foxcroft was a Fellow of King's College, Cambridge, and also (as some maintain) a principal mentor of Isaac Newton in the shadowy by-ways of alchemy. How much of the learning and beliefs of these people More took to himself it is difficult to determine. He was inclined to belittle his own competence in oriental tongues and rabbinical learning; his comment: 'I do not doubt but that there is pretious gold in this Cabbalisticall rubbish' bears a double edge (Nicolson 1930: 351, 5/2/1672).[19] But it is clear that he was fascinated by these mysterious thinkers.

For Van Helmont personally – a man of exactly his own age, though he was to outlive More by twelve years – Henry More grew to feel the warmest regard. (His feeling was shared by Lady Conway, who virtually placed herself under his direction, and left him a legacy of £300 in her Will, less than that to More of £400 indeed, describing both as her 'honoured and dear Friends'.) Departure from Ragley Hall in March 1671 provoked in More an outburst of emotion, as he later reported to his lady:

> [Van Helmont] has a hearte so good, so kind, so officious, so plaine and simple, and so desirous of the publick good, that the consideration of that in conjunction with something els, putt me into such a passion of joy and benignity, that I could not for my life keep my eyes from letting down teares, that morning in the parlour upon my converse with Mr Helmont, which not being able to suppresse, I took occasion to go up stares, into my Lords chamber to take leave of him, thinking to rid myself of this passion by the way, but when I came into his chamber, the more I endeavour'd to suppresse it the more it broke out, as old happinesse sometimes touches laughter in Melancholy men ... to compensate the losse of that moisture I spent at my eyes, my Lord sent for a can of Norden ayle ... After this his Lordship carry'd me down againe into the parlour and expressed his kindness to me in a glass of Canary, and we were then pretty humoursomely merry, and I excused myself as well as I could to Mr Helmont for that unexpected passion, and told him he was a Chymist and could draw moisture out of a flint. Such was the pass time of that morning, like sunshine and raine on an April day in February till as I told your Ladiship at last I prevail'd to gett away and take horse ... (Nicolson 1930: 329–30, 14/3/71)

This passage was written barely six months after More's first encounter with Van Helmont. On a day in October 1670 More, in his chamber in the New Building at Christ's College,[20] had introduced van Helmont to his friend Ezekiel Foxcroft, 'whose curiosity I thought it would gratify to converse with van Helmont, they both haveing a genius to Chymistry'. This was at the commencement of the mystic philosopher's long residence in England. He spoke both French and Italian but only broken Latin (so says More, perhaps meaning that van Helmont used the alien pronunciation of the continent), so that Foxcroft found it easier to understand him in his native Netherlands tongue. Van Helmont 'seems to be a person of very good plaine and experte humour, and yett with a dew guard upon himself. He does not profess Physick but lives on his own earnings,' More told Lady Conway. Later he was to see much of van Helmont at Ragley Hall, where they performed hydrostatical experiments together. Such esoteric questions as the nature of the apocalypse, the pre-existence of souls and cabalism were favourite topics of discussion between them (1930: 317–20; 323, 13/10/1670).

During his long stay at Ragley Lady Conway adopted van Helmont as her medical and spiritual director. To him the Quakers whom she gathered about herself were quite acceptable, whereas to More their religious and social principles were as detestable as they were absurd. Lady Conway died at Ragley Hall in February 1679, More's last visit there having been paid in the summer of 1677. During the final, completely bedridden stage of her sad life Lady Conway had found faith and comfort in the worship and customs of the Quakers, and (much to her husband's disgust) came in the end to choose only members of the Society of Friends as her personal attendants. Lord Conway wrote to his brother-in-law, Sir George Rawdon (December 1677): 'In my family all the women about my wife, and most of the rest, are Quakers and Mons Van Helmont is governor of that flock, an unpleasing sort of people, silent, sullen and of a removed conversation ... These and all of that Society have free access to my wife, but I believe Dr More, though he was in the house all the last summer, did not see her above twice or thrice, (1930: 439, December 1677). For her sake, More too pursued an acquaintance with the Friends, without ever softening his opinion of their 'enthusiasm' and melancholy. In August 1674 he entertained the much-imprisoned Scots Quaker, George Keith – who also met Lady Conway at about the same time – for between nine and twelve hours of discourse.

Keith was a learned man, an orientalist, and despite 'the ridiculous rusticity of that sect' More found him as an individual 'very considerably learned, of good witt, and quick apprehension, and which is best of all, heartily breathing after the attainment of the new life of the Christian. He is very philosophically and platonically given, and is pleased with the Notion of the Spirit of Nature' (1930: 391–2, 11/8/74). Later, More also met William Penn, the brothers George and John Whitehead, and Isaac Pennington. Though he found certain Quaker ideas totally unacceptable he was able to assure Lady Conway that 'There are some things which I hugely like in the Quakers and I wish all Christians would imitate them [there] in' (1930: 418, 10/1/1676).

After the cessation of the correspondence with Lady Conway More's life largely disappears from history; the extant, late correspondence contains nothing personal. His last published works were purely religious: an exposition, verse by verse, of the Apocalypse of St John the Divine (1680) and another of the prophecies of Daniel (1681), followed by a work on the eucharist in which he criticized the doctrine of transubstantiation (1681).[21] In September 1682 More presided over the obsequies of his friend Sir Thomas Baines, who had died at Constantinople a twelvemonth before: Sir John Finch had brought the embalmed body back for burial at Christ's. The Foundress's Chamber in the College was completely refurbished for this solemn occasion, and the grand dinner cost £16 17s. 3d. (Nicolson 1930: 461).[22] Thereafter More steadily declined: 'He was taken one Night after Supper very Ill in the Fellows Room, and swooned away: He complained afterwards "That his Distemper was Wind, but he hoped it would not carry him away in a Storm"' (Ward 1710: 214). From this attack he recovered, but in May 1687 Locke received an unfeeling letter from his friend John Freke including the following passage:

> Your conceit that Dr H Ms [spiritual] vehicles are of the race of Fairy pleases me much, but I doubt they are displeased with him for searching soe much into their affairs and prying too curiously into their world for his head has sufferd by it he has for severall months been deprived of the exercise of his great reason and had a Just title to a fellowship in the College called Bedlam but at present is pretty well recovered. (Locke 1976, III: 199)

Only a couple of months later Locke again had news of More from Damaris Cudworth:

> D:M: who is indeed very ill in Body and also Melancholy, as He Always was; though it now dispose him to be more unsociable, He not Careing for the Conversation of Any but his best Friends ... He has an Impatience (which I think indeed below so much Philosophy) to be in that Happyer State he has so well Describ'd; the Thoughts of which, and those other Great Truths it is Built upon, being the only support He says that the Mind of Man can have under so much Bodily indisposition ... (Locke 1976: 237, 29/7/1687)

Henry More entered that 'Happyer State' on 1 September 1687 and like John Finch and Thomas Baines, whose joint epitaph he had composed, was interred in the College Chapel. One recalls his writing to Lady Conway twenty years before: 'this life is a Pilgrimage, and I have a strong Presage that I shall finde myself amongst my more domestick friends when I am out of it, who will heartily congratulate my returne home' (Nicolson 1930: 200, 5/4/1662).

The personal bequests made by More in his Will, drawn up some fifteen months before his death, are not without biographical interest. There are scientific instruments left to his colleagues at Christ's: Cudworth received a 'parabolicall Glass', Thomas Standish (p. 89) 'my Ivory Globe Frame and case', and Thomas Lovett (c. 1645–93, Fellow 1667) 'all my Hydrostaticall Instruments'. Presumably More possessed no mathematical or astronomical instruments of value. Then there was a treasured portrait of Lady Anne Conway by the miniaturist John Hoskins (d. 1664), in a gold case; this had been presented to him, I believe, in late 1653 and was the gift made with 'so great and unexpected a measure of liberality and generous gratitude ... double above any imagination of mine' which More meant to keep as 'monuments of your goodnesse and generosity' (1930: 91). This relic he gave to Lady Conway's nephew, Henry Finch (b.c.1659), one of the first Fellows on the new Finch and Baines Foundation (1682). The other portrait mentioned, of himself drawn in black lead by David Loggan and engraved for his *Philosophical Works*, he gave to his sister's son, the Reverend Christopher Coleby. His former pupil, Dr Clark of Grantham, was left two medical books, the bulk of More's library going to Mr Coleby. Funeral rings were to be presented to the Master and Fellows of Christ's College in his memory, and also to a few named individuals among whom were his friend Edward Fowler, John Moore (later Bishop of Ely) and Isaac Newton, Lucasian Professor of Mathematics at Cambridge (1930: 481–3).

Part II

For and Against
the Scientific Revolution

6

Henry More's Philosophy

LIKE OTHER CAMBRIDGE PLATONISTS, Henry More was a follower of Plato and the neo-Platonists in a somewhat weak and general sense. His specific quotations from Plato are few. Plotinos he loved. His attachment to Hermes Trismegistus was restricted, at least until he fell under the spell of Francis Mercury van Helmont. He was not drawn to magic or alchemy; his deep distrust of 'enthusiasm' – might one render this word as 'mystical, incoherent raving without rational content'? – kept his feet on the ground, unless borne aloft by the fancies that he himself generated. More was very much his own man.

As we shall see later, the *Psychodia Platonica* seem to contain more echoes of Copernicus and Galileo than of Renaissance Platonism. The subjects of the four poems of 1642 are conventional and unspecific:

1   *Psychozoia*, or a Christian Platonicall display of Life
2   *Psychoathanasia Platonica*, or A Platonicall Poem of the Immortality of Souls, especially Man's Soul
3   *Antipsychopannychia*, A Confutation of the sleep of the Soul after Death
4   *Antimonopsychia*, or that All Souls are not one

Who in 1642 would stand forth to oppose such ideas? The most striking passage in the Platonism of this book is to be found in the Preface, where More traces an analogy (if not an identity) between the Platonic Triad and the Christian Trinity. He seems to say – for the passage is far from clear – that the *Ahad* [The One], *Aeon* [Eternity] and *Psyche* [Soul] of the Platonists are perceptions of God the Father, God the Son and God the Holy Ghost, respectively. The Platonist Principles are all 'omnipresent in the world, after the most perfect way that humane reason can conceive of. For they are in the world all totally and at once everywhere', as is God.

> This is the Famous Platonicall Triad: which though they that slight the Christian Trinity do take for a figment; yet I think it is no contemptible argument, that the Platonists, the best and divinest of Philosophers, and the Christians, the best of all that do profess religion, do both concur that there is a Trinity.

The next step in the reasoning of this passage seems to be incoherent since More, who might be expected to justify his preoccupation with the Triad by its analogy to the Trinity, seems instead to argue in the reverse direction:

> the Platonists placing him [that is, the second Unity in the Platonicall Triad] in the same order, and giving him the like attributes, with the Person of the Sonne in Christianity, it is nothing harsh for me to take occasion from hence to sing a while the true Christian Autocalon [ideal beauty], whose beauty shall adorn the whole Earth in good time. (1642: Preface to 'Psychozoia')

Fortunately, perhaps, there seems to be no return to these ideas in More's later prose works.

After his reading of Descartes, More's thought was dominated for many years by a different triad: Moses, Plato and Descartes. In these three names were summarized for him the whole history of philosophy, and of course the union of the pagan and the Christian traditions (Moses being a Christian *avant la lettre*). A good example of More's strange syncretism is to be found in a single sentence of the Preface to his *Philosophical Writings*:

> But that which enravishes me the most is, that we [Descartes and himself] both setting out from the same *Lists*, though taking severall ways, the one travelling on the lower *Road* of *Democritism*, amidst the thick dust of Atoms and flying particles of *Matter*; the other tracing it over the high and airey

Hills of *Platonism*, in that more thin and subtil Region of *Immateriality*, meet together notwithstanding at last (and certainly not without a Providence) at the same Goale, namely at the entrance of the holy Bible, dedicating our joint Labours to the use and glory of the Christian Church. (1662: Preface)

Here, while embracing Descartes as a philosophical brother, More already distinguishes the materialism of Descartes from the immateriality of Platonism. He seems to say that these are but two faces of the same coin, or perhaps object and reflected image, but precisely how the reconciliation between the two aspects of Janus-nature was to be effected he does not say, and it seems that any precise resolution of this problem for ever escaped him: that the world is God-made, that it is material and that it is infused with spirit he firmly believed, and therefore he was confident that the triple teaching of Moses, Plato and Descartes was capable of homogenization into one philosophy. His own efforts to 'reconstruct' it do not go beyond restatements of the historical theory of the *prisci theologi* and an awkward juxtaposition of disparate elements: 'the ancient *Pythagorick*, or *Judaick Cabbala* did consist of what we now call *Platonism* and *Cartesianism*, the latter being as it were the *Body*, the other the *Soul* of that Philosophy.' These becoming divided in antiquity, the metaphysicians went into excessive subtleties and 'ridiculous falsities concerning immaterial Beings', while Atomical Philosophers became 'over credulous of the powers of Matter'. 'I do not a little please myself, that I have made some progress towards resuscitating that *ancient and venerable Wisdome* againe to life, and the bringing together, as it were, of the *Soul* and *Body* of Moses' (1662: Preface, xviii). The difficulty here is with the metaphor: if we do not understand the union of body and soul in man, how are we to understand it in the doctrine of philosophy? For further enlightenment we must follow More's direction to investigate the Mosaic philosophy, the first and truest, recorded in the Holy Book. Not surprisingly, More finds in it his own favourite ideas, however misunderstood before: 'I do not know any one *Theorem* in all Natural Philosophy that has more sufficient reasons for it than the *Motion* of the *Earth*, notwithstanding [it] is part of the *Philosophick Cabbala*, or *Traditions* of Moses ...' when the Scriptural text is properly interpreted. 'So likewise for the *Praeexistency of the Soul*, which seems to have been part of the same *Tradition*, it is abundantly consentaneous to *Reason*.' The eternal life of the soul and the Trinitarian nature of the Deity are other Mosaic conceptions, according to More in his *Conjectura Cabbalistica, or a Conjectural Essay of interpreting the Mind of Moses* (1653b as reprinted in 1662: ii).[1]

The idea of the Trinity passed from Moses to the Platonists, who were in this respect (and others) his pupils. So especially was Pythagoras, who acted as an intermediary between the Eastern wisdom of Moses and the later Platonists:

> that *Pythagoras* was acquainted with the *Mosaical* or *Jewish* Philosophy, there is ample Testimony of it in Writers; as of Aristobulus, an Egyptian Jew in Clemens Alexandrinus ... St Ambrose adds, that he was a *Jew* himself. And though he gives no belief to the report yet that learned Antiquary Mr Selden seems inclinable enough to think it true in his first book *De jure Naturali juxta Hebraeos* ... Besides all these, Iamblichos also affirms that he lived at Sidon, his native country, where he fell acquainted with the Prophets, and successors of one Mochus the Physiologer ... Wherefore it is very plain that Pythagoras had his Philosophy from Moses. (1662: 110, 'Appendix to the Defence of the Philosophick Cabbala')

Another passage ties together Moses, Pythagoras and the motion of the Earth, with the difference that this time the informant of Pythagoras was not Mochus/Moses himself but a 'Jewish Priest' (or a Sidonian):

> it is well known, that the Pythagoreans held the *Motion of the Earth about the Sun*, which is plainly implied, according to the *Philosophick Cabbala* of this *Fourth* day's work ... it is very highly probable that he [Pythagoras] had the whole *Philosophick Cabbala* of the Creation opened to him by some knowing Priest or Philosopher (as we now call them) in the Oriental parts, that under this *mystery of numbers* set out to him the choicest and most precious Conclusions in *Natural Philosophy*. (1662: 82, 'Conjectura Cabbalistica')

Moses, Pythagoras and Descartes come together when More attributes to the first-named the origin of the idea, 'largely demonstrated in Descartes his Philosophy', that the Earth, in the form of a dead star, was created on the Fourth Day of the Creation, hence the importance of the number ten $(1 + 2 + 3 + 4 = 10)$.

Another major Platonic idea deriving from Moses (according to More) is that the union of the soul with mortal flesh represents its sleep; the true life of the soul is in that freedom from the body that men call death. Miracles such as Moses worked were not unknown in the Pythagorean records; indeed, the continuity from Moses and Pythagoras to Plato and his followers was complete, 'their Philosophy being the same that Pythagoras's was, and so alike applicable to Moses his text' (1662: 111).

In natural philosophy too More regarded the Pythagoreans as being the founders of Greek atomism. However, he supposed the origins of this theory to be older than Greek philosophy: for Mochus the Sidonian, already mentioned, who lived before the Trojan War if we may believe Strabo (and why should More doubt him?), and who was identified by Isaac Casaubon (and More) with Moses, was

> the first Authour of the *Atomick Philosophy*, or of that Philosophy that gives an account of the *Phenomena* from the figures and motions of the *Particles*. Whence there must be no small affinity betwixt this ancient Moschical, or rather Mosaical Physiology, and the Cartesian Philosophy, which has so often and so naturally born a part in this Philosophick Cabbala. (1662: 112; 113)

More believed that Diogenes Laertius confirmed the atomistic character of the Pythagorean philosophy, which was in turn transmitted to Democritos, who had studied with the Pythagoreans and became an intimate friend of the important Pythagorean, Philolaos. The link is supported, More thinks, by the story that when Plato proposed to burn the writings of Democritos, it was two Pythagoreans, Amyclas and Cleinias, who dissuaded him from his mistaken zeal.

Why then was it that Plato, following Pythagoras' philosophy upon so many points, rejected or at least passed in silence over the role of atomic composition in explaining the properties of things? For whether or not we may call Plato (on the evidence of *Timaeus*) a mathematical atomist, it is clear that he saw no reason to admire or even discuss the natural philosophy of atoms developed by Leucippos and Democritos. The Greek atomists' denial of a divine creator, and of the existence of soul as distinct from mind, the materialism of their philosophy and their assertion of bodily sensation as the ground of knowledge, were all anathema to Plato and every Platonist. Accordingly, it is unremarkable that the revival of Greek atomism by Renaissance authors (chiefly on the basis of Lucretius' *De rerum natura*), of whom Pierre Gassendi was in the seventeenth century the chief representative, was repugnant to More.

Descartes, on the other hand, More initially applauded as the reviver of Pythagorean atomism, precisely because he was *not* an atomist in the materialist tradition of Democritos-Lucretius-Gassendi. (That More misconceived the nature of Descartes's departure from this tradition is irrelevant at this point, though it was the reason for his later hostility to Descartes.) 'I cannot but surmise [More wrote], That he [Descartes] has happily and unexpectedly light upon that which will prove a true Restoration of that part of the *Mosaical* Philosophy, which is ordinarily called *Natural*, and in which *Pythagoras* may be justly deemed to have had no small insight' (1662: 112). According to More, we may deduce certain characteristics of the Pythagorean natural philosophy from the teaching that Democritos took from him:

> *Pythagoras* must have held, if Democritos had all his Philosophy from his Writings, or Traditions, That there are infinite Worlds, and that they are generable and corruptible, but that Matter [that is, atoms] is unperishable. That there are infinite numbers of Atoms or Particles ... and that *they are moved in the Universe after the manner of Vortices*. (1662: 113)

We observe that on all these points, whether or not More accurately reflected the ideas of Pythagoras and Democritos, he certainly reported the content of Descartes's *Principles of Philosophy*. Indeed, in More's eyes the great merit of Descartes was that he had restored the Mosaic-Pythagorean philosophy, and so the reader should take no offence at More's admission 'of the *Cartesian Philosophy* into the present *Cabbala*. The Principles, and the more notorious Conclusions thereof, offering themselves so freely, and unaffectedly, and so aptly and fitly taking their place in the Text, that I know not how, with Judgement and Conscience, to keep them out' (1662: 112).

As More and his friend Ralph Cudworth interpreted the history of philosophy, the overconfident Greek atomists (Democritos and Epicuros) had renounced the religious connotations of pre-Greek atomism in favour of a materialist philosophy. Therefore Plato, perceiving this, had rejected Democritos' doctrines and preserved in his own philosophy the true significance of religion and the spiritual element in Nature. Descartes had (as

More supposed) rightly abandoned the outright materialism of Democritos, but had not yet seen the full truth of Platonic spiritualist idealism; he was not as yet a Platonic atomist. It would therefore be a task for More himself to complete Descartes's resuscitation of the true philosophy by synthesizing it with Platonism. So a purified atomism and religion would be reunited as in the days of Moses. These notions, particularly More's idiosyncratic understanding of Descartes in his middle years, must seem strange to us, yet More's claim for the essential continuity of ancient philosophy was (as we have seen) perfectly familiar to the seventeenth century, as it had been to the sixteenth. More presents another version of the idea of a 'perennial philosophy' formulated by the Platonist Agostino Steuco in a book of 1540 which More may well have read. Another variant of the same idea inspired *The Court of the Gentiles* (1669–77) by More's contemporary Theophilus Gale (Schmitt 1966).[2]

The common statement that Henry More was an atomist whose teaching Newton followed needs qualification to the extent that More taught only *Platonic* atomism. He never approved the atomist philosophy of Epicuros revived in his own time; the title of his poem, *Democritus Platonissans* (1646), explains itself: it is the atomist who is to be amended. If More was to leave problematic the philosophical nature of his wished-for synthesis between Platonism and atomism, he was at least confident of its historical justification and the possibility of achieving it was encouraged by Descartes's example. More was not merely looking back to a Golden Age of philosophy in the days of Pythagoras and Plato. His view was that an esoteric but essentially correct world-view had become lost or distorted through the years during which other philosophical systems had prevailed, a world-view to which Descartes had returned, adding his own enrichments. He had also, defying Pythagorean prohibitions, opened it to all. (So also, when Newton later acknowledged his belief that Pythagoras had known something of the true system of the world which the *Principia* had restored, he was far from denying his own originality in writing that book, or from attributing to Pythagoras more than an inkling of a concept that the *Principia* had explored in rich detail.) The *prisci theologi*, it seems to me, were understood not as omniscient sages but as precursors whom the moderns had surpassed. In this way the idea of past wisdom and the idea of progress became complementary rather than antithetic; it is on the basis of the true principles grasped by the founders of science that progressive achievement can occur.

More envisaged a chain of 'true' philosophers: Mochus/Moses, the Pythagoreans, Descartes (coming after the long Aristotelian interregnum). After Pythagoras, Plato and his successors had developed a tradition of philosophy whose spiritualist metaphysics was admirable, but whose neglect of the material structure of the universe had limited its usefulness in natural philosophy. The parallel school founded by Democritos, on the other hand, while developing the natural-philosophical ideas derived from Moses and Pythagoras, had lost its way in atheistic materialism. Cudworth was to make much of this latter point. More was willing, when the evidence allowed, to fit other philosophers into the line that had re-emerged with Descartes, for example: 'something I have culled out in the life of *Parmenides*, that is so perfectly agreeable to the *Cartesian Philosophy*, that nothing can be more, and is indeed the very heart and marrow of it, and in a manner comprehends or takes hold of all' (1662: xviii, Preface).

The antithesis between the idealism of Plato and the materialism of the post-Pythagorean atomist tradition always affected More's attitude towards Descartes. For many years More tried to detach from atomism the concomitant notion of the entire self-sufficiency of matter, given to it by Democritos. But already in *Conjectura Cabbalistica* (1653b) he recognized that it was '*Plato*'s School, who makes the *Matter* unmoveable of itself; which is most reasonable ... for it is manifest, that there must be something besides the *Matter*, either to bind it or to move it' (1662: 78, 'Defence of the Philosophick Cabbala'). In truth, no principle could be less consonant to Cartesianism. More's increasing emphasis upon the spiritual power in Nature, bursting out in *Divine Dialogues* (1668), stemmed from a growing awareness that Platonism and Cartesianism could not be completely reconciled. In the collected *Philosophical Writings* of 1662 and especially in the *Epistola ad V.C.* there printed for the first time (it was to be issued separately two years later), despite the constantly reiterated praise of Descartes's natural philosophy, there are strong reservations. The Preface contains a forceful defence of the Hylarchic Principle: 'That there is a *Spirit of Nature*, that is to say, a substance incorporeal, that does interest it self in the bringing about some more general *Phaenomena* in the World, I think I have demonstrated so evidently, that nothing can be more evident in Philosophy.' This spirit is 'the lowest substantial *Activity* from the All-wise God, containing in it certain general Modes and Laws of Nature, for the good of the Universe. But the eye of particular Providence is not therein' (1662: xv;

xvi).[3] The Spirit governs all regular events of Nature: it gives matter activity and animation. It is not equivalent to the Soul of the World of the ancients, a notion More prides himself on utterly refuting. It is

> A substance incorporeal, but without Sense and Animadversion, pervading the whole Matter of the Universe, and exercising a plastical [formative] power therein according to the sundry predispositions and occasions in the parts it works upon, raising such Phaenomena in the World, by directing the parts of the Matter and their Motions, as cannot be resolved into meer Mechanical powers. (More 1659: 450)

Under God, it is an 'Inferiour Soule of the World' (p. 459) ensuring that consistency which God had determined to confer upon his creation but it does not act out of mechanical necessity. It is at once the cause of gravity and of resonance between musical strings, it shapes the foetus in the womb and makes flowers spring from the ground, because it contains within itself the general modes and laws of Nature (1659: ch. 12). By deliberately separating the concept of Law of Nature from its context of mechanical necessitarianism – the idea that each Law holds because it is impossible from the mechanical constitution of the world (created by God) that things should be otherwise – More was not so much disrupting the axiomatic structure of seventeenth-century physical science as removing necessitarianism from the axioms and Laws. Philosophers find by their investigations that these are confirmed because the Spirit of Nature, not blind necessity nor mechanical causation, upholds them. 'The only thing mechanical about the spirit of nature is that it acts regularly in certain instances, e.g. gravity, and this force is said to act in predictable ways in its interaction with matter' (Boylan 1980: 397). Though not capricious in its effects, the Spirit of Nature would certainly at the divine fiat be the cause of events counter to the ordinary course of Nature, such as the upheaving of the ground in earthquakes.

The Spirit is the cause of all natural motion. Where the mechanists argued that although the fundamental particles of matter are perfectly hard, the corpuscles composing common matter possess inherent elasticity, More maintained that the Spirit of Nature supplied the elastic repulsion between the corpuscles, which inert matter could not generate. The mechanists argued that a body immersed in a fluid of greater density than itself rises to the surface and floats there because the heavier fluid exerts an upward pressure upon it; More taught that the Spirit of Nature preserves the right order of things in which the light rests upon the heavy. From this and other examples it may be inferred that More's concept owed something to the Aristotelian organismic idea of the Universe. R. A. Greene quoted from the *Philosophical Poems* (1647) a passage that makes the parallel closer, 'Spermaticall Power' (that is, creative force) being the antecedent of the Spirit of Nature: '[the Spermaticall Power] belongs properly to Plants, but is transferred also to the Plasticall power in Animals. I enlarge it to all magnetick Power whatsoever ... For all magnetick power is founded in *Physics* [Nature], or else in *Psyche* [Mind]' (1962: 453–5). In broad terms, the Spirit of Nature is an agent preserving the integrity and harmony of the natural creation, which chance, necessity and the mechanical succession of events could not preserve.

Not unexpectedly – since More's Spirit of Nature has much the same function as an *explicans* as 'Nature' itself in the natural philosophy of Aristotle and Galen – More's physics, after his revulsion from Descartes, looked back to that of the Peripatetics. What is Order, which the Spirit of Nature preserves? It is the state in which things are in their right places in an hierarchical universe, and natural motion is the restoration of things to these rightful places:

> the *Spirit of Nature*, which ranges all things in their due order, acts proportionately strongly to reduce them thereto, as they are more heterogeniously [*sic*] and disproportionately placed as to their consistencies. And therefore by how much more crass and solid a body is above that in which it is placed, by so much the Spirit of Nature uses to reduce it to its right place ...

As in the pre-Galilean account of the acceleration of falling bodies, the greater the density of the descending body in proportion to that of the medium through which it descends, the greater the force causing it to descend. And when More in this passage goes on to explain why a bucket plunged under water feels progressively heavier as we raise it clear of the surface, again he echoes Aristotle: the Spirit of Nature resists our effort to remove an element out of its proper elementary sphere (Glanvill 1682: anonymous annotations by Henry More to ch. 14, p. 131). In the *Immortality of the Soul* (1659: Preface) More calls the Spirit of Nature 'the vicarious power of God upon Matter', and admits that it may seem 'to hinder that expected progress that may be made in the Mechanick Philosophy'. Yet Descartes himself had admitted that matter could not move at the beginning itself; it was necessary for God to superadd 'an impress of Motion upon it, in such a measure and proportion to all of it'. Unguided particles of matter by themselves would be quite unable to constitute organized bodies, *a fortiori* living creatures, any more than 'so many men blind and dumb from their nativity should joyn their forces and wits together to build a Castle' (More 1659: Preface; 67; 83; 87). 'Wherefore the ordinary Phoenomena of Nature being guided according to the most Exquisite Wisdom imaginable, it is plain that they are not the effects of the meer motion of Matter, but of some Immateriall Principle' (1659: 88).[4]

More's arguments, broadly speaking, are tantamount to the contention that such features of the natural world as consistency, order and repetition cannot arise by chance. In his eyes – as indeed in those of other critics – Descartes had failed satisfactorily to explain so universal and fundamental a problem as the cohesion of matter: to say that particles cohere because they are adjacent and at rest seemed a mere prevarication. To account for such attributes of matter as its cohesion, its gravity and its motion, More preferred to invoke the Spirit of Nature; so also with atmospheric pressure (p. 187). Adopting the Cartesian definition of inertia, More very fairly asks: what is it that prevents stones flung upwards from being 'carried away out of our sight into the farthest parts of the Aire, if some power more than Mechanical did not curbe that Motion, and force them downwards towards the Earth' (1659: Preface). If *mechanical* means what is comprehended within the laws of motion of Descartes (or of Newton) then it is true that no mechanical explanation of gravity can be deduced from them; auxiliary, particular assumptions have to be introduced (the theory of the celestial vortices for Descartes, the theory of universal gravitation for Newton). More strikes out directly for his own different solution: 'some immaterial cause, such as we call the *Spirit of Nature*, or *inferiour Soule of the World*, that must direct the *Aetherial* particles to act upon these grosser Bodies to drive them towards the Earth' (1659: 459). The vortex, no longer self-sufficient, is truly the immediate cause of gravity in More's view, but the direction of the vortex comes from the Spirit of Nature. Or again, as he had written earlier in *An Antidote to Atheism* (1653), the stars would not be round were it not for 'some Higher Principle than the mere Agitation of the Matter. But whether simply *Spermatical* [creative, as in plants] or *Sensitive* also, I'll leave to the Disquisition of others' (1659: Book III, ch. 12; 1662: 'Antidote against Atheism', 39ff., slightly abbreviated from the 1653 edition).

When he thus appeals to Spirit in explanation of the way the world works, why in other words a series of events A, B, C, ... consistently produces the result R, More is in the position of the simple animist who believes that his gods control the fruiting of trees and the flowing of springs, except that More's Spirit is neither wilful nor capricious and indeed does its work independently of the wills of men: the stone must always fall, the mercury stand high in the barometer. In its regular consistency of operation (for events A, B, C, ... never yield the result Q) the Spirit of Nature has to be the equivalent of the Cartesian mechanical system.

Why did More reject the concept of the self-enforcing Law of Nature, in the face of the authority of Descartes and of the objections brought against him by such empirical philosophers as Boyle and Hooke? The concept of Law was not irreligious, for philosophers were united in regarding Laws of Nature as fiats of the Divine Will. The difficulty for More was in the supposed obedience of inert matter to the Laws, as though it were conscious of them and could govern its own actions. To give matter a power of activity — of exerting an upward pressure or an elastic expansion or a magnetic attraction — was to transcend its limitation of inertness. More insists that the obedience of matter to the Laws of Nature must have an intelligent cause. Since it is unthinkable that matter is intelligent, then the cause must lie in the direct action of God or else in his indirect action through an intelligent agent. Newton was to prefer the first alternative; More firmly adopted the second. Perhaps for More the idea of God's acting continuously and universally upon every particle of matter in order to uphold the constancy of the universe savoured too much of continuous creation, contrary to the doctrine of *Genesis*.

More's attempts to defend his metaphysics against the mechanists were misguided. As we shall see, his criticisms of the new concepts of pneumatics embraced by Pascal, Boyle and Hooke joined him with an unenviable company, with whom he seemed to reject the whole idea of law-like behaviour in Nature. His attempts to defend the Spirit of Nature by appeal to the 'empirical evidence' for the operations of demons, ghosts and witches were a failure, irrelevant to his greater purpose of demonstrating God's perpetual activity in Nature. Later writers – including Newton – were to develop the arguments for God's far-sighted fabrication of the universe without appealing either to popular superstition or to any equivalent of More's Hylarchic Principle.

Nevertheless, More may be reckoned the pioneer, even the founder, of the English version of natural religion, a topic of little interest to the earlier Platonists of Emmanuel College. It is surprising that in his book on *Natural Religion and Christian Theology* (1953) C. E. Raven did not pay more attention to him. Several pages are devoted to Cudworth, his concept of 'plastick Nature', and his influence upon later theologians, yet there is no indication that the essence of all that Cudworth wrote had been expressed by More long before *The True Intellectual System* saw the light (1953: ch. 6). In this matter of the Spirit of Nature, as in others where there is a coincidence of thinking between the two Christ's men, priority in publication undoubtedly belongs to More. It may be true, as J. A. Passmore has argued in his study of Cudworth, that it is impossible to disentangle their intellectual contributions, but Passmore's argument in favour of his assertion that Cudworth was not only 'the leading systematic thinker among the Cambridge Platonists' but also (despite being by four years More's junior) 'very likely More's teacher' seems to me quite unsound. Cudworth's is generally supposed to be the stronger and more methodical mind, and this need not be disputed here, where the issue is rather: who was the more *original* thinker? Passmore argues from More's compliance with Cudworth's demands over the question of publishing a book on ethics, in 1665. Here Cudworth, with an unfinished work, was clearly the aggressor, in that he wished to stop the printing of More's book, which was ready for the press. Passmore supposes that More would not have yielded to Cudworth's demand had he not felt himself in a weak position, as an author indebted to Cudworth (1951: 18). But this is to make a question of fact turn upon a psychological hypothesis. Might it not have been the case that More gave way to his importunate friend – albeit only temporarily – even though equity in no way required him to do so? It seems hard (and unsafe) to read More's generous gesture as an indication of intellectual dependence. The positive evidence is all to the effect that More was not only more active in publication than Cudworth, and more capricious in his interests, but that he was also the more inventive thinker.

At any rate, as early as 1653 *An Antidote against Atheism* contains More's carefully detailed natural-philosophical arguments in justification of the concept that a benevolent Providence has shaped the world. If the Earth's constant motion about the Sun, the perpetual parallelism of its axis and the restorative force of gravity imply the existence of a knowing Principle to preserve order, such things as the distinction of land from sea on our planet, the form and beauty of living things as well as their usefulnes to man (often made known to him by their 'signatures'), the contrivance of the inner parts of animals – all these things demonstrate design and the conscious *creation* of order. So also such animals as the dog, the horse and the sheep have been fitted to serve man's needs, and the wonders of generation exist in order to continue the species of living things. More expatiates upon the intricate system of the heart and blood vessels, and especially upon that organ so often admired by his successors in natural theology:

> for the *Use* of the *Eye*, which is *Sight*, it is evident that this *Organ* is so exquisitely framed for that purpose, that not the least curiosity can be added... If the wit of Man had been to contrive this *Organ* for himself, what could he have possibly excogitated more accurate? (1712: 79, Antidote against Atheism)

So also, More continues, with the anatomical structure of muscles and of the ear. It is hardly surprising to find that John Ray, whose *Wisdom of God* (1691) was to be the finest seventeenth-century elucidation in English of the doctrines of natural theology, lectured upon *An Antidote* when he was a Fellow of Trinity College (Raven 1953: 118).

As a self-styled rationalist, Henry More put himself in much the same position as the Emmanuel Platonists. The first of the 'Rules' that he proposed under the heading 'How a man is to behave himself in this Rational and Philosophical Age for the gaining men to or the retaining them in the Christian Faith' reads:

> He must be sure not to deny any thing which he whom he would convince does hold and alledge upon clear and solid Reason: And especially he must be tender of denying it as repugnant to the Christian Faith; Unless it be plainly and really contrary to the Infallible Oracles of Holy Writ.[5]

That is to say, religious authority is no effective counter to secular reason. However, in the spirit of the prevailing opinion of his time that there are Divine Oracles open to human inspection, More adds in the Fifth Rule that no absolute reality and truth should be allowed to rational theorems; the philosopher should be 'more wary and reserved in the Assertion thereof' (Gabbey 1982: 227).

Though first published in 1664, these 'rules' seem to codify principles followed by More from the start of his career in writing, as Alan Gabbey remarks (1982: 228); Jentsch too found no development in More (1935: 60). Elsewhere in the *Apology of Dr Henry More* of which the 'Rules' form a part, More expands what he understands by this *'Seculum Philosophicum,* a Searching, Inquisitive, Rational and Philosophical Age, ... foreseen many and many Ages agoe by the Prophet Daniel'; it is an age in which philosophy may turn to the prejudice of the Christian religion. It is not difficult to think of the names of the chief philosophers at whom More might have pointed: Hobbes and Descartes. Early in life, therefore, More goes on 'I did set myself seriously and freely to search also into the most rational grounds of all such Philosophical Speculations as could any way pretend to have any moment for either the corroborating or enervating any Principle of Faith, or what Truths are recorded in the Holy Scriptures.' After long, diligent and anxious inquiry – the results of which are, by implication, expressed by More's previously printed works – he reached the conclusion dear to all exponents of rational theology: 'I can *ex animo* avow to all the world, *that there is no real clashing at all betwixt any genuine Point of Christianity and what true Philosophy and right Reason does determine or allow,* but that ... there is a perpetual peace and agreement betwixt Truth & Truth' (Gabbey 1982: 225–6). Certainly a comforting reflection.

As Gabbey goes on to say, More here declares overtly what is obvious enough and entirely within the spirit of Cambridge Platonism: his own primary role as a Christian apologist. The object of knowledge is religious fulfilment, for oneself and one's fellow men. The scholar's zeal for philosophy must have the justification that thereby he confutes the atheist and the scoffer; this is a view wholly in the neo-Platonic tradition and it conforms to the 'perennial philosophy' as stated by Agostino Steuco: 'the aim of philosophy is the knowledge of God, and, as it were, the actual beholding of Him' (Schmitt 1966: 520). Thus it would be quite in keeping with More's background to see his early writings as preparations for the explicitly theological studies he published from 1660 onwards . . . this was an interpretation he suggested himself, *ex post facto.* Yet one wonders whether the eager and pious young man of the 1630s, captivated by the new science of Copernicus and Galileo, could have looked so far ahead to a time when he would be ripe enough to turn theologian? Was not the satisfaction of those curious early philosophical poems in the writing of them, rather than in any ulterior purpose they might serve? Had he not declared: 'I would know that I may know?'

There was, of course, a change in the balance of More's literary output, a change commented upon by himself and recorded by Richard Ward who saw More as moving from natural to revealed religion. It is a change that followed the start of his disillusionment with Descartes, which may be dated to the middle 1650s, and his growing spiritualism. But clearly disillusionment with the philosophy of Descartes was not the cause of the change, though whatever provoked the former may have occasioned the latter also. Descartes had never been so important to More that the realization of his anti-religious tendencies could force More into a theological fervour formerly strange to him; nor were his literary energies from the late 1650s onwards devoted wholly to theology and the exposure of Descartes's errors; in fact a great deal of them were spent on the revision and translation Latin of his earlier philosophical writings. It seems that some deeper psychological or spiritual cause is involved, and this may perhaps be found in his relationship with Lady Conway. She was a far more religious and mystical philosopher than was More himself. Though of course the beginning of their friendship antedates even the first of More's major philosophical works, *An Antidote to Atheism,* the maturity of their intellectual intimacy coincides with More's shift from philosophy to religion. It may be that the preoccupations of her mind pushed his further down the road it might have been inclined to take in any case, for certainly she appealed to him increasingly, through the years, for guidance in matters of belief and worship rather than in philosophical questions.

The Platonism of Henry More, like his English style in verse and prose, has been variously esteemed; we may hope that recent attempts to interpret it have been more authentic as well as more sympathetic than those of earlier historians of philosophy, for whom More was a figure more than lightly tinged by absurdity and inconsistency. The older writers preferred the massive solidity of Cudworth to the fanciful flights of More, without sufficiently considering that the ideas of the former had been largely anticipated by the latter. Tulloch, for example, has little patience for the attachment of either scholar to the theory of the *prisci theologi*, neatly put in More's verses from *Philosophical Poems*, 1647:

> So if what's consonant to Plato's school
> (Which well agrees with learned Pythagore,
> Egyptian Trismegist, and th'antique roll
> Of Chaldee wisdome, all which time hath tore,
> But Plato and deep Plotin did restore),
> Which is my scope, I sing out lustily ...
>
> (Tulloch 1874: 314)[6]

To Tulloch's mind, any exposition of the ancient notion of the continuity of philosophy through the Greeks from its inception among the sages of Egypt and Asia, not to say of the once commonplace juxtaposition of the classical and Hebraeo-Christian mythologies, seemed 'labour thrown away, unless to illustrate the weakness of human genius or the singular absurdities which beset the progress of knowledge' (1874: 351–2). Recognizing more clearly the universality of such scholarly notions in the post-Renaissance period, and perhaps more conscious too of the real debt of Greek civilization to its imperial predecessors – of which indeed the memory had been preserved in the legendary histories of the Greeks – we now recognize that the idea of ancient wisdom was one of *the* great intellectual principles of the Renaissance, a principle which, though false in every detail, was surely correct in its historical spirit. More was, after all, right to believe that European philosophy and science did not have their most ancient antecedents in Hellas.

On the other hand, modern writers have rediscovered the exposure of the false antiquity of the Hermetic texts by Isaac Casaubon in 1614, and know that both More and Cudworth accepted Casaubon's demonstration, however reluctantly. The effect of Casaubon's exposure has not been examined in detail: conspicuously, it did *not* have the effect of destroying the tradition of the *prisci theologi*, nor did it even reduce the status of Hermes Trismegistus as an ancient pioneer of learning (even if less hoary with age than before). Still less did any scholarly discoveries concerning the dates of writers or the authenticity of texts weaken the appeal of Platonism as the most generous, the most Christian, the most ancient and the most true of the traditions of philosophy. Therefore it is hardly surprising that in a review of the 'intellectual and philosophical background' to the age of Milton, Samuel I. Mintz devotes the greater part of his essay to Platonism, and more pages to the English critics of Descartes and Hobbes than to these outstanding philosophers. Rightly, in my opinion, he distinguishes the Christ's Platonists from the Emmanuel men because they 'set themselves the larger task of consciously articulating a philosophical response to what they perceived as challenges to the view of a numinous, ordered, spiritual world' (Patrides and Waddington 1980: 158).

Their role, then, was defensive, with Descartes and Hobbes as the enemies. Recent authors such as Lichtenstein, Mintz and Gabbey are less puzzled by More's change of attitude to Descartes than was Tulloch. More had charged Descartes with driving God out of the universe: 'and many other [such] objections More insists upon at length, with an acerbity and exaggeration of statement no less remarkable than his previous raptures. His Cartesian enthusiasm has at length not only died down, but turned into sourness and bitterness of spirit' (Tulloch 1874: 368ff; 376). Tulloch recognized, of course, that More had never abandoned Platonism in favour of Cartesianism, and so his devotion to the Cartesian type of mechanical philosophy had been less than entire, as also that More could never accept the Cartesian treatment of animals as machines: 'He was altogether of a different temper of mind.' The strange thing is not that More should have repented his partial approval of Descartes, but that he should ever have expressed it. As he was to declare in 1669: 'the Mechanical Part of the Cartesian Philosophy, which boasts that all Phenomena in the World can be explained only by motion and substance . . . and that there is no extension not material or corporeal, most sweetly misleads scholiasts and half-educated men' (Patrides and Waddington 1980: 160). Does this mean that in the 1640s More himself was among those thus misled? Not entirely: as we shall see, More always understood that the world of mechanism was purely material, and he welcomed Descartes's sketch of such a world because it indicated the limits of materialism. The world of matter (as Descartes described it) was insufficient to be the real world. In More's reading, Descartes had indicated not how much mere materialism could achieve, but how *little*; and the rest of Nature was the realm of spirit. It was when More finally saw that the mechanists meant to claim that their world was not a part of the real universe but the whole of it that he came to see them not as friends but as foes. The dispute, after all, is less about Cartesian physics than Cartesian metaphysics, less about the philosophy of mechanism than the philosophy of science. Like so many after him, More was posing the question: 'Where are the limits of science?' and narrowing them down to his own satisfaction. Moreover when More, in effect, accuses Descartes of making God an anomaly within his own creation, he is surely putting his finger on a real difficulty in Descartes's philosophy: that God must *be*, for epistemological reasons, but cannot *do*, for metaphysical reasons. How different from More's own contention that 'if the Power of God be somewhere, God is somewhere, namely where the Divine Power is; *He* is in every part of *Matter*, the *Soul* in the humane *Body*' (More 1671; Mackinnon 1925: 188).

So also with More as a rationalist, a difficult point with the Cambridge Platonists, as we have seen. Like his fellows, More taught that reason was the only safe guide for the philosophical theologian. But he plainly declared also that the fruit of reason comes not from reasoning but from the reasoner:

> All pretenders to philosophy will be ready to magnify Reason to the Skies, and to make it the light of heaven, and the very Oracle of God; but they do not consider that the Oracle of God is not to be heard but in His holy temple − that is to say, in a good and holy man, thoroughly sanctified in spirit, soul, and body. (Tulloch 1874: 356, his translation from the Preface to More's *Opera Omnia* (1679))

This, of course, is sound neo-Platonism. But Tulloch, perhaps forgetting the ancient and damnable association between Platonism and magic, asks why the rationalist More 'developed so largely, not only an element of mysticism, but a vein of credulity, which must be pronounced excessive even for his age' (1874: 358). It is strange that More's credulity has more often received adverse comment than that of Joseph Glanvill or Dr Thomas Browne or Dr Robert Plot, first Curator of the Ashmolean Museum in Oxford, who inserts into the list of questions about natural history and antiquities upon which his county histories were to be based the inquiry: 'What apparitions have been seen?' Should More have been *less* credulous than most of his contemporaries (His Majesty's judges among them) concerning ghosts and witches? Again, the odd point is not that More should share the ordinary belief of his time, as that he should exploit this folk-wisdom to philosophical effect (as Glanvill did too). Perhaps it would be more just to indicate a certain naivety of mind in More, rather than blame him for a foolish tendency to superstition. The truth is that what people believe to be fact or falsehood has little to do with reason or rationalism as these words are used by philosophers; their decisions are based on authority and evidence at best, on the popular currency of information at worst. Authority and evidence, as well as popular wisdom and the bent of his own thoughts, assured More that there were good and evil spirits at work in the world.

Moreover, as Mintz has emphasized, the Cambridge Platonists' 'holy temple' requirement for the exercise of right reason lifted their beliefs above the mundane world of logic, evidence and the balance of probabilities. In the phrase of John Smith: 'That which springs forth from true goodness ... brings such a Divine light into the Soul, as is more clear and convincing than any Demonstration' (Mintz 1962: 82, quoting Smith's *Discourses* (1673)). This is so because the pure soul is an image of God; so Henry More: 'The Soul of Man is as it were *agalma theon*, a compendious Statue of the Deity; her Substance is a *solid Effigies of God*' (More 1653a; MacKinnon 1925: 40). In the good man intuition and reason are one; he knows the truth not merely by calculation, as it were, but by illumination. For More the fact of God entailed the fact of spirits, therefore it was nothing less than rational that such evidence of spirits' action as Tedworth provided should be presented to him. 'No spirit, no God'.

A fascinating example of the complete acceptability of More's position in his own time, and of his contemporaries' thorough understanding of it, is provided by the 'American' plays of John Dryden in which free use is made of supernatural beings, aerial and earthly spirits and demons. Literary scholars agree that the dramatist's presentation of these beings is probably based on 'the writings of Henry More and Joseph Glanvill, both of whom shared [Dryden's] need to rediscover spiritual forces within a world increasingly politicised and materialistic' (Armistead 1986: 140). Hogarth too, in a later generation, was well aware of the association between the 'enthusiasms' of witchcraft and of madness, while the early followers of the Wesleys regarded insanity as a manifestation of demonic possession.[7]

A final statement of the place of Henry More in English intellectual history can hardly be made, because he is a man of paradox. Everyone agrees that he is an immensely attractive figure, though many lament that his works are unreadable. That scarcely a year passes without the publication of a book or major article in which More is given a prominent position suggests that he is still read, if only by academics out of duty. In his own time, according to a well-known story, 'the *Mystery of Godliness* and Dr More's other Works ruled all the Booksellers in London,' and the number of reprints and fresh editions of them down to modern times has not been small (Lichtenstein 1962: 13). He was certainly a contributor to what Lichtenstein has called 'the movement for the minimizing of both dogma and intellection in religious life, and the emphasis upon simplicity and practical morality' (1962: 30). (Note here again the antithesis, attributed to More, to rationalism in a Thomist sense.) Richard Baxter regarded him as the chief of the Cambridge Platonists. Henry More seems, indeed, to be one of those historical personages about whom a little is widely known; who is frequently upon the stage, if not often under the spotlight; who has figured in the background of a Victorian historical romance (Shorthouse 1881); who appears in every dictionary and work of reference with a respectable entry; but of whom it can still be written: 'The imprint left by More upon the shores of seventeenth-century intellectual history the waves of time have since partially effaced ... The tribute of recognition is one which Henry More richly deserves, and it has been long overdue' (Lichtenstein 1962: Preface).

The fact is that Henry More has never been unrecognized, anonymous, forgotten, though indeed he may have been misunderstood and regarded as more absurd than he really was. But we must be content to leave him as a minor figure of intellectual history. For he is a strange, idiosyncratic and very human author. Often seemingly inconsistent in his metaphysics and his world-view, his combination of grandiose imagination, of mechanism and of spiritualism, his peculiar alloy of subtlety and naivety of thought, all place More in a class by himself. He had no successor.

7

The Spirit World

In the learned discussions of the seventeenth century, the protean term *spirit* signified a principle or substance that is active, elusive, impalpable, fugitive and mysterious. Chemists called a volatile fluid such as alcohol a spirit; physiologists found three spirits (natural, vital, animal) within human beings governing their physical and mental faculties; the Holy Spirit was the most abstract member of the Trinity; an ordinary spirit might also be a subordinate supernatural being, an angel or a demon; the spirit of man or woman was the soul, but also a supernatural apparition or ghost when the human spirit was manifest detached from the natural body; the Spirit of Nature or soul of the world (the vital force of later centuries) was an invisible, active agency everywhere present in Nature, the antithesis of matter animating matter, giving it motion and in animals (significant word!) life. There are of course conceptual and etymological connections between all these senses of the same word, and of partial synonyms like 'soul', 'ghost' or 'essence'.

In Henry More's philosophy the most important senses of the word *spirit* are as signifying the human soul or the soul of Nature.[1] Of course these are very different things. The one identifies an individual human being, the other is generalized (for More did not imagine each horse or tree to possess its own soul, as a person does); but these spirits are alike in being substantial (though their substance is penetrable, impalpable and continuous) and immortal. These are the spirits of the ordinary course of human and natural existence. But More was also much interested in extraordinary or supernatural manifestations of the real and independent existence of spirits: the ghosts of the dead, or angels or demons. For although well aware of the philosophical distinctions between involving the concept of spirit in the ordinary, recurrent affairs or phenomena of humanity and Nature, and invoking the same concept to account for the supernatural mysteries such as witchcraft, More believed that the one use of the concept gave strength to the other. The invisible presence of normal spirits was confirmed by the visible manifestation of ghost or demon. Indisputable testimony to the appearances of ghosts and similar supernatural events (as More believed) served to demonstrate the existence of immortal human souls, just as witchcraft demonstrated the existence of the Devil. If no Devil, then no God.

Without departing from ideas fundamental to his thinking from first to last, More in the latter part of his life progressed steadily deeper into esoteric and spiritualist associations. The first flush of his enthusiasm for Descartes's mechanical philosophy – with which he was never in total agreement – proved short-lived, and he came to think that though this philosophy was valid within certain limits, the sphere within which it was invalid (for all the endeavours to the contrary of Descartes and his successors) was very wide; within this sphere spirit ruled and matter was its servant. The assertion of the primacy of spirit over matter was firmly made by More in his *Antidote against Atheism* (1653a), where he declared that the very notion of matter is more 'unconceivable than anything in the notion of spirit. For my owne part, I think the *nature* of a Spirit is as conceivable and easy to be defined as the nature of anything else.' To prove his point, More then defined spirit as a substance 'possessing Self-penetration, Self-motion, Self-contraction and dilatation and Indivisibility ... Penetrating, Moving and altering the Matter' (MacKinnon 1925: 11; 12). At this time and for many years thereafter, through such works as *The Immortality of the Soul* and *Divine Dialogues*, More proclaimed his doctrine of spirits in a metaphysical and religious context. So far as his spirits had a role in scientific explanation it was in relation to phenomena that are admittedly highly perplexing, such as those of gravity, or else in relation to living creatures. More became a vehement opponent of Descartes's mechanistic physiology: 'We have now particularly evinced [he declared] that neither the whole

Body [of Man], nor any of those parts that have been pitched upon, if we exclude the presence of a Soul or Immaterial Substance, can be the Seat of Common Sense.' And again, in a more precise claim, 'the chief Seat of the Soule, where she perceives all objects ... is those purer Animal Spirits in the fourth Ventricle of the Brain' (1659: 180; 197). The main business of the material organs, the brain and nerves, was to serve as the vehicles for these spirits in order to enable them to act within the body. When republishing *An Antidote against Atheism* in 1662, however, More took occasion to attack in Book II the 'materialist' theory of the spring and weight of the air, recently promulgated by Robert Boyle. More insisted that only a substance distinct from matter could so act as to cause the effects Boyle had described (see below, chapter 9). As he assured Lady Conway: 'I be not altogether satisfyde that his paradoxicall Inferences from the experiments are true. There will be a Spirit of Nature for all this, and I think for anything that ever will be alledg'd to the contrary, or excogitated to evade the relish of that principle' (Nicolson 1930: 269, 17/3/1666).[2]

The historian may feel that however hard it was for Boyle to prove that the atmosphere is elastic, it was an even stiffer task for More to prove the existence of an immaterial, undetectable spirit pressing the air into empty bottles. The Spirit of Nature could be invoked in order to evade any mechanical hypothesis, but how to grasp so insubstantial an entity? Hence the importance of those other inhabitants of the spiritual world, the spirits of the departed which were only thinly veiled from mortal eyes. Ghosts were inhabitants of the spiritual world of this planet – into which, More believed, all the dead enter, if only for a time – become visible to human eyes. More and his friends avidly collected stories of their appearance. But spirits (as modern lore confirms) are more often heard than seen. There is a story, repeated by Marjorie Nicolson without good authority (and it seems to be a fiction) that More in 1670 investigated 'psychic' phenomena at the ruins of the Priory of Cookhill, not far from Ragley Hall. An ancient cottage had been constructed in the ruins; through a stone slab in its floor strange murmurings, sighings and calling were heard, and ghosts were said to appear and vanish through it. More had the slab raised and, finding steps below, boldly descended with a candle. A passage led him to a great vault where he found skulls and bones and the tomb of Lady Isabel de Beauchamp, also a plain elm coffin less than a century old. Dripping water and whistling draughts might, More thought, explain the sounds that troubled the cottagers (1930: 249–50). The desire to investigate such an affair seems typical of More (though he was usually content to swallow ghost stories on trust); the sceptical search for matter-of-fact explanations is not. He would certainly have thought it wholly impious to try to invoke demons by magical procedures and apparently made no experiments with talismans or other mild, one might say Christianized, forms of magic. As with other scholars of the age, including Newton and Leibniz, the Hermetic tradition was of great interest to More even though he stands outside it. In general his attitude to all these matters was that of the contemplative, literary scholar and not at all that of the active magician.

The principal supernatural phenomenon that came to More's notice in his middle years – at second or third hand, it is true – was that of the Drummer of Tedworth, Wiltshire, in 1661. (The Mompesson family was persecuted by a drummer to whom insufficient alms had been given, with loud drummings and other noises for which no human cause could be found.) To a modern reader these events closely resemble those stories of persecution by a *poltergeist*, recorded by Harry Price and other modern students of the occult; one may plausibly believe that the Mompesson phenomena were the work of human cunning and malevolence. At the time (1663) they seemed to many, and above all to Joseph Glanvill who published them with great gusto, the firmest evidence for the material activity of spirits that had been manifest for many years, and they were welcomed accordingly as weapons against scepticism.

The enlarged or 'More' editions of *Saducismus Triumphatus* (1681, 1688, see below) contained an epistle from More to Glanvill urging the latter to print the fuller account of the Tedworth visitation which indeed appeared in them, in order to refute those who decry the story as an imposture, and to buttress this with 'other Stories sufficiently fresh and very well attested and certain'. Because of Glanvill's early death More himself was to undertake this task. More's biographer Richard Ward found nothing extraordinary in More's conviction of the reality of 'ye story of ye *Daimon* of Tedworth' which he took to be

> a most remarkable Instance of this kind. It is surprising to consider, wth wt a strange Confidence, yt long and signally attested Narrative hath been run down by many; and wt a Rumour hath prevail'd yt both Mr *Mompesson* and Mr *Glanvil* confessed all *to be a Cheat and a Contrivance*. Never was there

any Thing more falsly, vainly or groundlessly spread about than this pretended and unaccountable Report. (The MS continuation of Ward's *Life*, in the
Library of Christ's College, Cambridge; Nicolson 1930: 254–5)

More appealed to the Mompesson case as evidence of the reality of spirits
against the doubts or 'sadducism' of Hylobares, a character in the *Divine
Dialogues* (1668) modelled upon Thomas Baines, bosom friend of John Finch.

John Finch – the best company in the world, according to More, though
strangers remarked his haughtiness – had adopted Baines as the inseparable
partner of his life since the two young men had met as undergraduates at
Christ's College, and More's pupils. Four years after the publication of
Divine Dialogues, in 1672, the pair bade farewell there to More, before
setting off for Constantinople, where Finch was to be Ambassador; after
their convivial evening together More wrote to Lady Conway of Baines
that though his 'legges faile him, yet his tong walkes as free as ever, and
he is very good company on that account, and really I phancy his minde
is of a better frame than heretofore', meaning by this that Baines was less
of a sceptic with respect to spirits and the supernatural (Nicolson 1930:
357).[3] At the time of the Tedworth drumming More had written to her:

> O that Dr Baines that witleather wit had but had the hap to have been at
> Tedworth in those transactions, What tough tugging there would have been
> betwixt the evidence of sense and the prejudicate phancyes that his Taurine
> Blood had hatched against the existence of Daemons. But this Areall [aerial]
> Drummer would have beat a retreat to all his forward conclusions of that
> kinde. (1930: 216, 31/3/1663)

If not in real life, at any rate in the *Divine Dialogues* More ('Bathynous') was
able to force Hylobares from incredulity to a position of agnosticism or
grudging admission of the existence of spirits.

In later years More continued to express great interest in stories of the
supernatural. Early in 1674, for example, he promised to procure for Lady
Conway 'an authentic copy of a story Mr Francis Robarts told me here
[Cambridge] of an Apparition ... The most philosophicall one that I ever
heard of before' (1930: 385, 23/2/1674). And three years later he told her,
in relation to the deaths of two of his Irish pupils and their mother, her
relatives: 'I suspect those dancings of the 3 lights on Fairy Mont in Ireland

was ominous and presag'd the death of the three persons lately deceased' (1930: 431, 29/4/1677).[4] Whether or not Lady Conway actively encouraged More in the trend to spiritualism in natural philosophy that accompanied his flight from Descartes, she certainly contributed such relations as these to his repertoire of supernatural phenomena.

The trend was one that More and his coadjutor, Joseph Glanvill, felt to be fully in the tradition of Plato. More demonstrates a certain caution in fathering upon the Platonists his own preoccupation with ghosts and witches, but this caution was not shared by Glanvill who, in effect, builds the Principle of Plenitude into the Platonic philosophy. All possible combinations and manifestations of matter and spirit together must actually exist, he asserts, and so there must be delicately material beings, immensely more spiritual than ourselves, intermediate between us and the angels, who are composed of pure spirit. Observe in this connection, says Glanvill:

> that I have again made use of the *Platonick Hypothesis, That Spirits are embodied*, upon which indeed a great part of my Discourse is grounded ... Since then the greatest part of the World consists of the *finer* portions of matter, and our own Souls are *immediately united* unto these, 'tis infinitely probable to conjecture, that the nearer Orders of Spirits are *vitally joined* to such *Bodies*; and so *Nature* by *degrees* ascending still by the more *refin'd* and *subtile matter*, gets at last to the pure *Noὲς*, or *immaterial minds*, which the *Platonists* made the *highest Order of created Beings*. (1688: 92)

Obviously Glanvill here writes of a variety of distinct entities, some of which might be angels, some demons.

More shared Glanvill's belief in the existence of such beings ('the Platonick Hypothesis, that Spirits are embodied') and used this belief to justify his Spirit of Nature (pp. 114 – 17) though this, a universal agent acting as an omnipresent vehicle of the Divine Will, is of a very different order. In his eagerness to prove the real existence of spiritual entities active in the universe, More allowed himself no discrimination in the evidence relating to them in their various categories of action. He finds no difficulty or incongruity in asking his readers to believe that the human soul is immortal and that the Moon is guided by a spirit, on the basis of stories about a suicidal shoemaker of Breslau or the bad behaviour of the Abbess Magdalena Crucia whose liaison with the devil lasted for thirty years. Books that begin

at the highest level in learned metaphysical discussion end with such a tale
as that of Johannes Ginochius 'who in the month of July saw with his own
eyes a drop of rain suddenly turned into a frog', or of Marsilio Ficino's
appearing on a white horse at the moment of his death to his friend Michael
Mercatus, crying out 'Michael, Michael, vera vera sunt illa'. 'The yarns are
great entertainment, and we can appreciate why More thought them
significant, but as we read them the Cartesian way of philosophizing
becomes a dim and distant melody' (More 1659: 272; 293; Gabbey 1982:
203). True; and what is worse, any hope of separating the philosophic
discussion of God in natural philosophy from old wives' tales and age-old
prejudice against sorcerers vanishes also.

More worked very hard to define precisely what he meant by spirit,
or 'immaterial substance', denying that the last two words constitute a
contradiction; spirit is substantial because it occupies space and is capable
of affecting matter, for example, by moving it. Spirit like matter is created –
only God is uncreated. In most respects spirit is the antithesis of matter.
Indeed, More insists that in principle if the attribute A exists, then not-A
must also exist, and vice-versa. So, since matter is not self-moving, something must exist that is self-moving, that is, spirit. Since matter is discerpible
(capable of being divided) and impenetrable, spirit is indiscerpible and
penetrable. The particles of matter are of fixed dimensions, but spiritual
substance can expand and contract indefinitely; and so forth. The human
soul is spirit and is the source of intellectual life, which can be the product
neither of the whole body nor of any single organ within it. The soul is
essential to the human body but not the body to the soul, which exists
before and after corporeal existence. Indeed, the soul may leave the living
body for the 'free Aire' and return to it; it can traverse 'solid Iron and
Marble as well as through the soft Air and Aether' (1659: 80; 197; 268;
283). Such a migration of the soul is as natural 'as how the Fire will ascend
upwards, or a Stone fall downwards for neither are the motions of these
meerly Mechanicall, but vital or Magicall, that cannot be resolved into meer
Matter.'

The pre-existence of both human and animal souls is 'more agreeable to
Reason than any other Hypothesis whatever' (1659: 275; 240). Souls not
incorporated in bodies, says More, occupy the air, along with angels and
demons. At least the good types of such spiritual beings 'may administer
much content to one another in mutual Conferences concerning the nature

of Things, whether Moral, Natural or Metaphysical ... But of all Pleasures, there are none that are comparable to those that proceed from their joynt exercise of Religion and Devotion.' More paints a charming picture of other aspects of this immaterial existence which, however, hardly transcends corporeal imagination:

> Wherefore they cannot but enravish one anothers Souls, while they are mutual Spectators of the perfect pulchritude of one anothers persons and comely carriage, of their graceful dancing, their melodious singing and playing, with accents so sweet and soft, as if we should imagine the Air here of itself to compose Lessons, and to send forth Musicall sounds without the help of any terrestrial Instrument ... (1659: 400; 421)

If these flights of fancy now oppress by their extravagance – though Dr Johnson admired them – More also devoted many pages to demonstrating, against the materialist physiology of Descartes, that the soul is the essential engine of life. The very existence of free will 'is a Demonstration that there is a faculty in us that is incompatible to meer Matter', as are all such 'Faculties or operations as we are conscious of in ourselves'. Perception, motion of our bodies *ad libitum*, memory and imagination cannot be performed by a mechanical system of pineal gland, muscles and fluids. 'Memory is in the soul, and not in the brain.' Spirits are the immediate cause of muscular contractions: 'That there is some fluid and subtile Matter, which we ordinarily call Spirits, directed into the Muscles that move the Member, its swelling does evidence to our sight.' The heart is the seat of the 'plastick' (we might say, the active or energizing) congruity of the soul with matter, the brain the seat of the perceptive congruity. Soul has all our perception, and hence all our pleasure (1659: 110; 154; 232; 264–5; 270, etc.).

This is what we might call the 'least claim' for the soul or spirits, preserving, as it does, the familiar distinction of organic from inorganic (mechanical) which Kant was to clarify in the next century. In maintaining such a distinction against Cartesian materialism, which reduced animals – and the human body – to the level of mechanisms, More was in harmony with the predominant feeling of his own and later times. While most natural philosophers and natural historians of the eighteenth century agreed that mechanical, physical and chemical explanations had their places in the analysis of living phenomena, few denied that something lay outside these explanations, some vital force or spirit that resisted reduction to materialism (see Raven 1953; Roger 1963; Roe 1981; Mazzolini and Roe 1986). But More was not a man to be satisfied with a least claim. Nor was he inclined merely to find spirit at work where the next generation of Newtonians would discern forces. In his eagerness to claim the organic from Descartes, he swept the inorganic into his own realm of spirits. The claim he chose to make was that everything in Nature is animated: even matter, so-called, possesses a dull kind of life that inspires it to motion.

In the last of the letters written to him by Henry More that Descartes read and answered (in draft only), dated 23 July 1649, More had argued that either rest or motion must be natural to matter, and that, whichever choice was made, it was impossible to understand how one moving particle (or body) could impart the same motion to another particle (or body) at rest. And how could motion drive the particle about, if it were so lightly attached to it?

> Hence [he continued] I am the more inclined to this opinion, that there is absolutely no transfer of motions; but that a second body is as it were awakened into motion by the impact of the first body, as this or that event awakens the soul to reflection. And that the second body does not so much receive motion from the first, as put itself into motion at the bidding of the first. And that as I have said just now it is the same with motion in a body as with thought in a mind, that is, neither is received but both arise within the subject in which they are found. And that all we call body has a dull and sottish life, though destitute of sensation and perception, as being the last and lowest shadow or image of that divine essence which I take to be the most perfect life.
>
> Moreover, your transfer of motion from body to body, from greater to less and less to greater, as I noted above, represents very well the nature of my extended spirits, which can contract themselves and expand again, easily penetrate matter without filling it, impell and move it any way you please, and yet without any instruments or the attachment of hooks. (Descartes 1974: 383; cf. Gabbey 1982: 211)

Here, it is clear, More has either failed to understand or has rejected the doctrine of motion developed from Copernicus to Galileo and Descartes. Suppose we compare the impact of one moving particle upon another with the handing of a baton from one runner to a second in a race. More denies that the transfer of the baton is the cause of the second man's starting to run; the cause of his running is his own act of will, occasioned by receipt of the baton. So with the particles: the movement of the particle which is struck is a new motion, of which matter is not the cause. To put this another way: the motion of the second particle after impact is caused by the elasticity of both, and elasticity for More is a spiritual effect. If we deny this, we are saying that matter can move itself. Further, it is evident that for More the only motion that is philosophically worthy of consideration is *absolute* motion; not for him the point later made so clear (and so heuristically useful) by Christiaan Huygens that it is only the observer's chosen frame of reference that determines which body rests and which is in motion, which is the striker and which the struck. How could it be (according to Huygenian relativism) that a body should not 'know' (in More's terms) whether it is moving or resting? More's distinction between a bidding and a bidden body could only be drawn in a world of absolute motions.

Descartes did not deign to criticize More's animism seriously. Dismissively, he wrote of More's drunken living matter as *suavia* (chat) and uttered a solemn warning: 'You will allow me to say, once for all, that nothing more impedes our discovery of truth, than positing those things as true for which no rational argument can be constructed, but which are supported only by our own volition' (1974: 405; Gabbey 1982: 212). A severe rebuke, not undeserved, which sufficiently indicates the difference between Descartes and More as philosophers. If Descartes meant that More philosophized about a world conceived in his imagination to suit his own tastes, he was right: but More might have riposted with a *tu quoque.*

When he read this draft reply by Descartes to his last received letter, in 1655, More was not at all abashed and in answer to the long-dead philosopher wrote (presumably for the benefit of Clerselier) a rejoinder that perhaps went further than it would have done six years earlier in asserting that the true existence of corporeal spirits was witnessed by the more than 600 true histories, no fables, testifying to their reality (Gabbey 1982: 213).

The unholy alliance between Henry More and Joseph Glanvill in defence and illustration of their belief in demonic activity and witchcraft was chiefly effective in an extraordinarily complex book, *Saducismus Triumphatus* (literally, *Agnosticism Overcome*). Written originally by Glanvill in 1666 under a different title, after its author's death the book was twice reissued (1681, 1688) by the bookseller S. Lownds, with extra matter added by

Henry More who thus became its co-author.[5] The essence of the whole work is contained in two propositions which it purports to demonstrate:

1   Active, immaterial spirits exist and are known to men;
2   Witchcraft and other forms of demonic activity are genuine occurrences providing indisputable evidence of the existence of spirits.

To Glanvill and More these propositions were not merely philosophically important; they were of the greatest theological weight also. In their eyes, he who denies spirits denies God. To doubt the truth of witchcraft was to advance more than halfway towards atheism. As More wrote in the Preface to the second edition of *Saducismus Triumphatus* (1681):

> That I am thus very industrious and zealous to support the belief of *Spirits* and *Apparitions*, and of whatever is true that contributes thereto, may seem strange to some, and therefore to want an Apology; yet considering the *Saducism* of this present Age, and Atheism too if you will, it were a great neglect in me, or any one else of my Profession, not to have a great zeal and indignation against the stupour and besottedness of the men of these times, that are so sunk into the dull sense of their Bodies, that they have lost all belief or conceit that there are any such things as Spirits in the World.

More goes on to say that the 'Wits' who deny witches, spirits and souls frustrate God's purpose in sending Christ upon earth and the agony of his crucifixion: so

> What real Christian would not be moved to the height of indignation, at so foul and frantick a Scene of things, and industriously lend his hand to the amending it as far as he can. That I should hope may be a just Apology for my thus zealously assisting, and faithfully ministring to the serious Design of our dear Lord and Saviour. (1688: 14–15)

In fact, as this passage implies, both the argument from witches to God and its inverse were used by the apologists, for they were equally sure that the reality of God and of the scriptural writings confirmed the existence of demonism. So More again: 'if there were any modesty left in mankind, the Histories of the Bible might abundantly assure men of the existence of

Angels and Spirits.' Conversely, More points to the errors of 'such course-grain'd Philosophers as those Hobbians and Spinozians, and the rest of that Rabble' who slight religion and the Bible precisely because of the stories of spirits and angelic visitations which these philosophers reject as 'things that their dull Souls conceit to be impossible'. Accordingly, More writes:

> I look upon it as a special piece of Providence that there are ever and Anon such fresh examples of Apparitions and Witchcrafts as may rub up and awaken their benummed and lethargic Minds into a suspicion, at least, if not assurance, that there are other intelligent Beings besides those that are clad in heavy Earth or Clay. (1688: 25; 26)

From this appears the reason why God permits witches to exist: the 'Confession of Witches against their own lives' — or in other more direct words the admissions extracted under torture — together with their miraculous feats being plain evidence that there are bad spirits, and so opening the door to the belief that there are good ones, and a God. To collect and record well-attested stories of witches and apparitions is not to cultivate folly and superstition but to 'do real service to true Religion and sound Philosophy, and the most effectual and accommodate to the confounding of Infidelity and Atheism, even in the Judgement of the Atheists themselves, who are as much afraid of the truth of these Stories as an Ape is of a Whip' (1688: 26–7). The assertion that scoffers at witchcraft are believers in their hearts is common in the literature of demonology.

Presumably More gathered his 'Stories' of ghostly visitations and demonism from the many earlier writers on these topics, but he does not seem often to refer to them by name. One whom he mentions approvingly, the considerable scholar Meric Casaubon (1599–1671), had assisted More's defence of spiritualism by several books, notably *Of Credulity and Incredulity against the Sadducism of the Times in denying Spirits, Witches etc.* (1668). More did not stress, as the witch-hunters did, either the universality of witchcraft or the vast harm it did. But, like them, he found the Biblical authority for his own opinions overwhelming (the Witch of Endor, studied at some length by More, must be cited in every book on demonology) and, like the witch-hunters, More dismissed the incredulous as atheists. The argument that scepticism concerning supernatural manifestations is equivalent to scepticism concerning God was perhaps new with More. With one exception he equally ignores writers who criticize the witch-hunters, though he must surely have known Reginald Scot's *Discoverie of Witchcraft* (1584). The exception is John Webster's *The Displaying of Supposed Witchcraft* (1677). John Webster (1610–82), a somewhat distant associate of Samuel Hartlib and a warmer enthusiast for the republican reform of England, was unlikely to commend himself to More. He was a chemical philosopher – a group of whom More had a low opinion – and he had attacked the established privileges of the English Universities. His *Academiarum Examen* (1654), 'the most extensive critique of scholastic education produced by the puritan reformers', had been answered by John Wilkins and Seth Ward (Webster 1975: 198–202). More had previously castigated Webster's scepticism in the Preface to his Latin collected works; in the later editions of *Saducismus Triumphatus* he printed a double disquisition addressed to Glanvill on the errors of Webster's treatment of the story of the Witch of Endor raising the ghost of Samuel for Saul. Webster's book was, in More's phrase, 'but a weak, impertinent piece of work, the very Master-piece thereof being so weak and impertinent, and falling so short of the scope he aims at ...'. Webster's assertion that 'there is nothing but Couzenage [deceit] and Melancholy [mental illness] in the whole business of the feats of Witches' was false, and it was untrue that accounts of the gross and obscene relations between witches and demons were credited by any of Webster's 'Witchmongers, as he rudely and slovenly calls that learned and serious person Dr *Casaubon* and the rest'. (Here More went too far: the 'witch-mongers' do indeed dwell with relish upon such stories.) 'But as for the true and adequate notion of a Witch or Wizzard ... his Arguments all of them are too weak or impertinent' (More 1688: 52; 53).

It might seem from More's words that he and Glanvill had set aside from their record the grosser kind of supernatural occurrences which critics of the belief in witchcraft had derided as absurd. Not so: they are to be found prominently in *Saducismus Triumphatus*, though indeed avoiding obscenity. To give one example in brief, the story of Anne Bishop, who persuaded another woman to perambulate the Church at Wincanton with her three times widdershins and at night:

> In the first round, they met a Man in black Cloths, who went round the
> second time with them, and then they met a thing in the shape of a great

black Toad, which leapt up against the examinant's Apron. In their third round they met somewhat in the shape of a Rat, which vanished away.

The witness, Alice Duke, under examination further related that she had compacted with the devil (who gave her sixpence) on a signed paper or parchment – a thing More had previously said no one believed. It is also recorded that Alice Duke's familiar took the shape of a cat – again, the kind of story of witch behaviour that More had accused Webster of mistakenly attributing to witchmongers (1688: 358–9; 364).

Many of the tales in *Saducismus Triumphatus* were collected by More and sent to Glanvill for use in his campaign. Two were borrowed from Robert Plot's *Natural History of Oxfordshire* (1677), one from a book on hydrostatics (1672) by George Sinclair, later author of *Satan's Invisible World Discovered* (1685). One account, of what happened in Ireland to a butler sent out to buy a pack of playing-cards, involved the Earl of Orrery and was related by Valentine Greatrakes at Lord Conway's table at Ragley. Other stories came to More from correspondents in various counties of England, among them such dignified friends in the Church as Edward Fowler, prebendary of Gloucester, and George Rust, Bishop of Dromore. A tale from Holland of a Dutch woman miraculously healed by an apparition in 1676 was 'procured by a friend from Dr Cudworth'; it was further authenticated by a printed broadsheet brought to Ragley by van Helmont and the testimony of Philip Limberg writing to Henry More. The haunting of a house in Bow, London, was described by a witness to More, who passed it on to Glanvill. The witness was Dr Gibbs, prebendary of Westminster, to whom More had been introduced by William Outram, also a Fellow of Christ's. A narrative by Andrew Paschall, sometime Fellow of Queens' College, Cambridge, of disturbances at his father's house in London in 1661, was found by More 'by chance in mine own Study'. Paschall told the same story to Cudworth, Fowler and Glanvill.

These relations of supernatural occurrences stretch in time from before the murder of the Duke of Buckingham by Felton in 1628 (foretold by the ghost of the Duke's father) to 1682. Evidently More's friends were assiduous, over many years, in reporting such events to him.

More was conscious of the necessity to authenticate his 'relations' as heavily as possible: a hundred doubtful stories of ghosts or witches are useless, one convincing example is enough. Therefore he was careful to invoke the names of peers and learned divines, if not as eye-witnesses yet as admitting the evidence for the occurrences. Some of these were taken from the Assizes of Cork and Somerset, and from trials in Scotland. Others came from Robert Hunt, a JP in the latter county who had 'a Book of Examinations of Witches, which he kept by him fairly written'. Many accounts came from more than one 'independent' source, as did the story concerning the Duke of Buckingham (More speaking): 'This story I heard … with great assurance, and with larger circumstances, from a Person of Honour, but I shall content my self to note only what I find in a Letter of Mr *Timothy Locket* of *Mongton* to Mr. *Glanvil*' (1688: 411). Another example of multiple transmission:

> This is a Copy of the Narrative sent by Mr. *Pool*, Oct.24 1672 to Mr *Archer* of *Emmanuel Colledge*, Nephew to the Judge [presiding over the trial of the witches] … But I remember here at *Cambridge*, I heard the main passage[s], of this narrative, when they were first spread abroad after the Assizes (1688: 392)

Two stories were vouched for by Lady Conway herself, in two letters from Ireland of 1662 and 1663, although on the report of other persons (1688: 461; Nicolson 1930: 214). Of one of the stories taken from Plot, More concludes that 'it is manifest that these Pranks were play'd by *Daemons*, as that learned Physician Dr *Willis*, a good many years ago did readily acknowledge to me, and avouched such things as here related in the Story to be undoubtedly true' (1688: 486) in the course of a conversation at Ragley. But Willis knew no more of the specific matter than More did. The story told by George Sinclair was similarly 'verified' by Gilbert Burnet, Bishop of Salisbury, 'who upon my enquiry told me this, That he living in Glasgow some years, found all People there and in the Country about, fully perswaded of the Truth of the Matter of Fact' (1688: 497–8).[6]

It was More's practice to seek such endorsement of the stories that he printed. In the preface to the second edition of *Saducismus Triumphatus* he prides himself on confessing an error in the first; in recounting the trial of Sharp and Walker at Durham Assizes, the recollections of a person had been included who had been no more than a boy at the time, not of full age, and another person had since admitted only hearing of something stated in the first edition to be of his own knowledge. But, More goes on naively enough, this solitary mistake in the matter of the whole being thus freely and ingenuously acknowledged, it will instead of a *dead Fly* in the Box of Ointment, prove only *Naevus in Venere,* a meer Beauty-spot in a Face of good feature and sound and fair Complexion. For there has been all moral diligence used, that nothing should be admitted that was any ways suspected, or exceptionable. (1688: 11)

Historians of Cambridge Platonism have singled out More for reprehension because of his 'credulity', a weak leaning towards popular superstition inconsistent with the high rationalism of a philosopher, that was not shared (as they suppose) by other leading members of the school, notably not by Ralph Cudworth. 'Credulity' and 'incredulity' are tendentious words, their use being wholly dependent on the intellectual temper of the user. More would not have been thought credulous by Shakespeare, Milton or Dryden; by the Bishops of his Church who supported his spiritualist endeavours; by any of His Majesty's Judges; or by a majority of the Fellows of the College of Physicians. It seems severe to condemn Henry More for embracing and proclaiming a belief – however repugnant to us – that was commonplace in his own time. We may *wish* that More's had been one of those rare minds that rejected the odious doctrine of witchcraft and distrusted (at least) the whole mythology of ghostly visitations, but that is a different matter. One might as well wish that Samuel Pepys could have consulted a modern ophthalmologist. An historical epoch must be accepted warts and all. Henry More without his ghosts and witches would not have been More.

Perhaps one may compare with the modern view of More's 'credulity' a recent interpretation of his spiritual metaphysics:

> More's arguments can perhaps best be viewed, from our present point of view, as an endeavour to reunite matter and spirit, which the rigid logic of Descartes had left in unbridgeable opposition, and to give greater 'body', or actuality, to both conceptions, which in Cartesianism were too nakedly abstract. More wants his 'spirit' to be more than abstract 'cogitation'; he will have it to be activity, and the activity must be *there where* it is at work, penetrating and moving matter. (Willey 1949: 167)

What is curious in More is his juxtaposition in his writings of the grand philosophical purpose analysed by Basil Willey, and the lofty notions by which More sought to implement it, with the mire of witchcraft trials. Was the clearest evidence of the power of spirit to penetrate and move matter to be found in an old woman's black cat or the tantrums of a poltergeist? Why should the existence of the virtuous spirit whose highest essence is in God be deduced from the evil spirit manifest in the works of the Devil? More's object was to show that the notion of spirit at work in the universe contained no inherent contradiction or absurdity, but his method seems completely deficient in logic, since it is to affirm that good is proved by evil, order by caprice. How (to insist yet again) is the idea of the Spirit of Nature, whose function is to maintain the beneficent system of the universe, strengthened by the contention that the universe is full of the capricious, contra-natural activity of an evil spirit?

More's illogicality has the effect of embarrassing natural philosophy with just the same antithesis between good and evil that had always embarrassed theology. For of course the absence of a coherent logical transition from More's metaphysical disquisitions of spirit to his 'relations' of supernatural occurrences has an exact analogue in the impossibility of passing smoothly from the philosopher's or the scientist's conception of God to the God of Christian belief. Consider, for example, Cudworth's definition (with which Newton's celebrated phrases in the General Scholium concluding the later editions of the *Principia* might be compared): 'The true and genuine idea of God in general is this, A perfect conscious understanding Being (or Mind) existing of itself from eternity, and the cause of all other things' (1845, I: 321). Now the particular tenets of any faith cannot conceivably be unpacked from such a definition. If the Christian theologian is asked upon what authority God may also be said to be loving or vengeful, to be Triune, to exercise care over the fall of a sparrow, to have created man in his own image, and so forth, the only answer he can give is that he knows these characteristics of God from the sacred writings of the Church, by the word of venerable authors and trustworthy witnesses and from his own inner conviction. More's case for witches and ghosts is exactly the same, and had a similar universality of honest belief. From the necessary idea of spirit he could, perhaps, deduce no more than the existence of both good and evil spirits, expressing the duality of moral nature. For the rest he depends upon the word of venerable authors, trustworthy witnesses and his own inner conviction. And in part the historical evidence to which he appealed was identical with that employed by any Christian apologist. The case for

More's 'credulous' beliefs is exactly the same as the case he would have made for his religion, as he was perfectly aware himself. If the one is defective both in metaphysics and in logic, resting ineluctably upon the contingent, so too is the other.

We can now see that already in More's generation the cultural tide was beginning to move against the belief in demonism, so that in this as in other respects More appears to us a backward-looking figure (Thomas 1971: ch. 18). In Britain, at least, the great days of witch-hunting were passing away and Christianity was slowly adjusting itself to the contemplation of a universe that was no longer the scene of an eternal battle between the invisible spiritual representatives of good and evil. But is not this struggle between God and Devil deeply rooted in the human unconscious mind? Willey is surely right in suggesting that More, like Milton, perceived the sharpening of the monotheistic sense of religion brought about by awareness of this cosmic drama. In using the crude weapon of popular superstition to defend religion, More and Glanvill may have been 'guided by a sound instinct':

> They may have obscurely felt, though they could not have realised or admitted it, that the ancient springs of popular demonology were those of religion itself, and that in the emotion of the supernatural, however evoked, they had a surer foundation for faith than all the 'proofs' of philosophic theism. (1949: 169)

Of course, if More had been able to realize and formulate to himself as an intellectual proposition any such analysis as this, he would inevitably have perceived also that it ran counter to the basic ideals of rational theology, thereby opening the door to 'enthusiasm'. More's whole position in the history of thought really rests upon his not possessing any such understanding of the antitheses in the human mind between belief and reason.

8

More and Descartes

WE KNOW THAT from the first Henry More rejected the greater part of the metaphysics and moral philosophy that he found in Descartes's writings; why then did he embrace and teach Descartes's natural philosophy? His deepest and surest loyalty was to his religion. Before reading Descartes he had plunged far into neo-Platonism, and he might easily have gone down the same path as Robert Fludd. He was never to be in any significant sense a natural philosopher himself, nor did he ever undertake an original investigation. On the other hand More, before he became acquainted with Cartesian philosophy, had been drawn to admire the mathematical account of Nature, which might have pre-disposed him to applaud Descartes's endeavour to extend the scope of reasoning *more geometrico*. If More failed to perceive at once the dangers inherent in Cartesian materialism, so readily obvious to other University men in France, the Netherlands and Britain only a little later, perhaps it was because he lacked their firmness in defending the established teachings of Aristotle and Galen, and so also lacked their inclination to employ the charge of atheism against Descartes, in defence of the *status quo*. Fundamentally conservative as he was, More was never satisfied that whatever is, is right. And since (as we have seen) More never ceased to teach Cartesian physics, perhaps his position was not far distant from that of John Locke, who held (1684) that while no system of natural philosophy could be 'taught to a

Young Man as a Science, wherein he may be sure to find Truth and Certainty, which is, what all Sciences give an Expectation of', nevertheless, 'the modern *Corpuscularians* talk, in most Things, more intelligibly than the Peripateticks, who possessed the Schools immediately before them' (Axtell 1968: 304–5, quoted by Rogers 1985: 296–7).[1] The corpuscularians who possessed the Schools were of course the followers of Descartes.

According to his recent biographer, Cudworth too adopted the corpuscularian philosophy pragmatically as the most plausible available; indeed, so great was his agreement with Descartes on so many vital issues that it 'is not misleading to call Cudworth a Cartesian' (Passmore 1951: 8; Saveson 1960: 560). Passmore is far from limiting this agreement to natural philosophy only, but he makes the point that, like More, Cudworth's attachment to dualism led him to embrace corpuscularianism: 'We can never sufficiently applaud that ancient atomical philosophy, so successfully revived of late by Cartesius, in that it shows distinctly what matter is.' Though Passmore was perhaps too readily inclined to identify Cambridge Platonism with Cartesianism, it seems to be true that Cudworth never reacted so strongly against it as More was to do, nor did he so positively embrace a spiritualist philosophy. He remained much where his friend had been in the 1650s. Of all the Cambridge Platonists only John Smith gave 'unqualified acceptance to Cartesian mechanism' (1951: 10; 20–1, quoting Cudworth's *Treatise Concerning Eternal and Immutable Morality*; 4, 6, 15).

If to be a Cartesian is to subscribe unreservedly to every word in Descartes's writings, then Henry More was never a Cartesian. This is John Rogers's view as it was Alexandre Koyré's (Koyré 1957: 110; Rogers 1985). Alternatively, if to accept Cartesian natural philosophy and to hold that Descartes's explanations of most of the phenomena of Nature are probably correct is to be a Cartesian, then it would seem that More never moved far from this position, at least before the last decade of his life. It might be argued that in his final years More rejected natural philosophy altogether – as certainly he had given up all hope of learning anything from experimental investigations – and put his trust uniquely in inner illumination and the doctrine of spirit it brought to him. If a man rejects not only Descartes and Boyle, but Aristotle and Galen too; if he denies the possibility of any analytical account of Nature and scouts all attempts to see a logical sequence of causation in things in any form; if he can find no action in Nature not immediately attributable to animism, then the question of his preferences in natural philosophy ceases to have meaning. He can have none.

However we describe the process, it is certain that More moved from a greater approval of Descartes to a less, indeed to a denunciation. When he first read Descartes's *Principia Philosophiae* (1644), More applauded its author as a Christian philosopher like himself:

> It seems ... that for More in 1646 Descartes was essentially a new and gifted (Christian) thinker in the Atomist tradition, the author of a mechanical cosmology, which, pruned of its refinements and separated from its philo-sophical foundations (which More in any case thought inessential), provided him with influential contemporary support in his magnification of the divine goodness and the soul's immortality, just as the teachings of the Atomists, purified of their darker aspects, provided him with an ancient pagan source for the justification of the same Christian truths. (Gabbey 1982: 182)

And so, according to a tradition for whose confirmation there is little enough contemporary evidence, More almost at once began to introduce Cartesian philosophy into his Cambridge teaching. It seems fair to judge that More was fitted neither by temperament nor by his studies to make a judicious appraisal of the technical merits of his hero's contributions to mathematical and natural philosophy: he does not spend time on comparing Descartes with Aristotle or Witello, nor does he enter into discussions of the differences between Galilean and Cartesian mechanics; the appeal of Descartes to More was of a wider and more ideological character:

> The philosophical and theological reasons for More's initial involvement with Descartes's philosophy are well known: Descartes's concern to establish the existence of God and the immortality of the soul, his revival and 'Christianizing' of ancient infinitist conceptions about the natural world, his belief in the Platonic ideal of the unity of knowledge, the clarity and distinctness of his philosophical approach, his rejection of substantial forms and the physics of the Scholastics in favour of a new and successful kind of natural philosophy. (1982: 190)

More's initial declaration in favour of Descartes was published only two years after the *Principia Philosophiae*, in More's *Democritus Platonissans: or an Essay upon the Infinitie of Worlds*, as Marjorie Nicolson discovered long ago (1929b: 362; 1930: 43–4). There is no evidence of More's having earlier read the *Discours sur la Méthode* – indeed, he knew little French (Gabbey 1977: Cahier 3, 12–13). The poem as a whole, however, is Platonic rather than Cartesian (the idea of an infinity of existence is Plato's, not Descartes's) and written with an ecstatic extravagance. The poet-philosopher who cries 'Then all the works of God with close embrace, / I dearly hug in my enlargèd arms!' is beyond caring for the world of clear and distinct ideas. It is not true that 'in the early period, there were no doubts in More's mind' about Descartes, for More never ceased to believe that Plato was a greater, richer philosopher.[2] More can write endless praise of Descartes, it is true, in this first phase of English Cartesianism, can speak of 'his excellent & transcendent wit', and declare that even his disciple the Princess Elizabeth of Bohemia (a queen of philosophers) possesses more wisdom than any man in Europe (Nicolson 1929b: 362–6). But he often recurs to the *limitation* of Descartes's intellectual ascendancy: he is a 'sublime and subtile Mechanick'. Or, as he put it elsewhere, Descartes has furnished the body of philosophy but its soul came long ago from Plato. As we shall see shortly, there was much in Descartes's philosophy that was deeply offensive to More's sensitive spirit.

In *The Immortality of the Soul* More urged the wide dissemination of Descartes's teaching:

> I think it is the most sober and faithful advice that can be offered to the Christian World, that they would encourage the reading of Descartes in all publick Schools or Universities. That the Students of Philosophy may be exercised in the just extent of the mechanical powers of Matter, how farre they will reach and where they fall short. Which will be the best assistance to Religion that Reason and the Knowledge of Nature can afford. (1659: 67)

Evidently this is something less than Descartes himself hoped for, and hints at More's reservations. At the same time, as we have seen, he was incorporating Descartes into a 'cabbalisticall' tradition extending from Moses, and in the *Defence of the Threefold Cabbala* printed in the collection of *Philosophical Writings* he confesses his admiration afresh:

> I should look upon *Des-Cartes* as a man most truly *inspired* in the knowledge of Nature, than any that have professed themselves so these sixteen hundred years; and being even ravished with admiration of his transcendent *Mech-*

— 149 —

anical inventions, for the salving the *Phaenomena* in the world, I should not stick to compare him with *Bezaliel* and *Aholiab* those skilful and cunning workers of the Tabernacle, who, as Moses testifies, were filled with the Spirit of God, and they were of an excellent understanding to find out all manner of curious works. (1662: 114; quoted in Rogers 1985: 293)

Again, whatever Descartes might have thought of this comparison with ancient (though holy) artificers, the praise is unmistakeable.

In this same work of 1662 More makes many specific allusions to the Cartesian theory of matter, which in his characteristically curious way he wishes to assimilate to *Genesis*; so, the agitation of matter at the creation

> brought it to *Des-Cartes* his second Principle, which is the true *Aether* ... *The Heavens* ... *are Fire and Water*; which no Philosophy makes good so well as the *Cartesian* ... the *Earth* consists of the third *Element* in the *Cartesian* Philosophy, (for the truth of that Philosophy will force itself in whether I will or no;) ...

> the *Celestial Matter* does consist of two plainly distinguishable parts, to wit, the *First Element*, and the *Second*, or the *Materia subtilissima*, and the *round Particles*, as I have already intimated out of Des-cartes his Philosophy. (1662: 78–9)

More apologizes to his reader for so frequently introducing Descartes's philosophy into his 'Cabbala', for the reason that his principles and the conclusions drawn from them offer themselves so freely and take their places so naturally in his text that he does not know how 'with Judgement and Conscience, to keep them out' (1662: 112).

In the *Epistola H. Mori ad V.C. ...* ('The Letter of H. More to Mr. ⸻, including an Apology for Descartes, which might be considered as a Substitute for an Introduction to the whole Cartesian Philosophy'), More asks himself: why does he study this philosophy with so much zeal and pleasure? Because it is mechanical, he answers, and everything in physical Nature is held to arise from motion modulated by the size, shape and position of material particles, while recognizing that it was God who first impressed motion upon the world. It is a grand thing to be able to philosophize from the immutable and necessary laws of Nature, and many phenomena can be demonstrated from these laws. But these do not amount to one in a thousand of the whole. It would be an absurd and superstitious adoration and worship of blind matter to attribute all phenomena to it; Descartes went too far in this direction. Descartes was an incomparable philosopher, and we may profit from his errors because they teach us the limits of materialism. For when effects are found in Nature which cannot be attributed to mechanical laws (and there are many such), we see the necessity for introducing an immaterial and corporeal principle, known as spirit. The colours of the rainbow and of the peacock did not arise by chance; there has to be a governing providence over all things.

Descartes may be criticized, More adds, because in explaining the phenomena of Nature he indulged excessively his genius in mathematics and mechanics. Thus his philosophy is not, after all, the product of unimpugnable reason and certain principles, but rather the consequence of Descartes's personal interests and his mentality![3]

From this time the flow of More's publications became steadily more hostile to Descartes. The full range of the technical difficulties he found in Descartes's books was made known to the learned in 1657, when Claude Clerselier published the first collection of Descartes's letters, including those written to him in 1648 and 1649 by Henry More, still near the high point of his enthusiasm for the French philosopher.[4] He had made the first approach to Descartes at the suggestion of Ralph Cudworth and Samuel Hartlib, the latter arranging the transmission of the correspondence. Inevitably More, whose name at this time must have been unknown to the recipient, opens his first letter (11 December 1648) with a page of encomium: his exultation in following Descartes's theorems can hardly have been less than that of the author in composing them; he had never expected to encounter such a close conjunction with his own ideas; other investigators of Nature, living and dead, seem like dwarves and pygmies compared to Descartes. Everything in his writings agrees so exquisitely with Nature itself that the human intellect could seek no higher satisfaction (Descartes 1974: 236–8).

After the soft soap, More's criticisms are by no means trivial or few, though (at the beginning) rather aimed at the generalities than the details of the system. Descartes admitted that if successfully pushed, More's objections would undo his work. More was to comment on several occasions that Descartes seemed to have adjusted his definition of motion to avoid contravening the Roman Church's condemnation of Copernicus, 'so that no man's Reason could make sense of it' while yet 'Modesty [did not] permit him to fancy anything Nonsense in so excellent an Author' (1662: Preface, xi; 1664a: 1). More imparted this difficulty to Descartes; he himself believed that time and space, rest and motion, were absolute attributes. This, and his belief that spirit is extended – that is to say, it occupies space – caused More further problems with Descartes's denial that a space void of matter can exist in the world. For Descartes, God, spirits and the human soul had no substantial existence in the universe. Further, More could not understand why Descartes refused to say that the universe is infinite (he called it 'indefinite') and also refused to allow that the least particles of matter are atoms. More wanted Descartes to say that matter is tangible and impenetrable as well as extended, but the French philosopher refused to give way on any of these points, all of which he regarded as essential components in the texture of his thought. But the doctrine of Descartes that most appalled More was his denial of life and sensation to animals – More calls it a sentence of execution, 'a deadly and lethal pronouncement', a sword cutting off the living stream from all creatures leaving them as mere statues or machines. How could parrots imitate us if they do not heed our speech? Does not a dog manifest guilt and shame, when he steals food? 'That copious set of stories by which not a few authors seek to demonstrate the presence of reason in dumb animals at least makes it clear that they possess sensation and memory' (Descartes 1974: 244). But More perceives the difficulty: in Descartes's terms, if an animal can think, it must have an immortal soul – *cogitat ergo est*. And that they have such a soul was, he adds, the Platonic opinion.

To Descartes's reply of 5 February 1649, which made no concession to More's objections, the English philosopher responded a month later, opening with his satisfaction that Descartes combined such courtesy and polished manners with his wonderful breadth of mind and divine acumen. (More's letters throughout, as printed, differ from Descartes's businesslike answers in being elaborate and flowery.) More's is an immense letter – twenty printed pages – responding to Descartes's answer point by point; the same metaphysical issues about extension, infinity, God and matter are traversed as before. At one point More implies that Descartes would have men play with shadows as cats play with their tails; at another he asks whether human babies have no souls during the first months of life since they behave like animals? He again denies that emendation of the Cartesian system to accommodate his own opinions would necessarily destroy it, and asks whether it would not be fitting for a philosopher to acknowledge a certain incorporeal substance in the world which might be the cause of all, or most, of the attributes of material particles such as shape, motion, and position, from which in turn arise perceptible properties like colours?

> Lastly, since incorporeal substance has this stupendous power (*virtus*) that by its mere application it can bind together or separate, divide and project or control matter without the use of strings or hooks, wedges or fastenings, may it not seem likely that it can contract itself together, since nothing impenetrable impedes it, and expand itself again, and the like? (1974: 315)

At the close of this letter More put to Descartes some more detailed questions about the structure of the aetherial and material particles which Descartes had imagined as responsible for most of the phenomena of physics.

In his second reply, postponed to mid-April 1649, Descartes again stood firm, point by point, repeating that the extension of matter did not apply to the human mind, to spirits or to God. He provided adroit answers to More's awkward conundrums: we know babies have minds, because adults have minds, but animals never grow up in this way; it is impossible to poke a stick beyond the edge of a finite universe because where the stick goes, there the universe is, and so on. All this time Descartes largely ignored the animism implied in More's allusions to his own preferred notions, to which More returned when he wrote a third time on 23 July asserting his own belief (along with the Platonists and the Fathers) that all souls and spirits, good and evil, can assume corporeal form and then possess sensation (1974: 377).[5] No one could read More's letters to Descartes without understanding that More from the first rejected mechanism as the foundation of philosophy, never budging from his initial Platonist conceptions that matter is inert, spirit the sole active agency in the natural world.

Only in his unfinished draft reply to More's third letter did Descartes allow himself to rebuke More for excessive preoccupation with such questions as corporeal angels and the shadow of the divine essence (see above, p. 137). The remainder of this short draft is concerned with More's difficulties about rest and motion: justifiably, Descartes accused More of failing to comprehend his idea of motion as a *modus*, and his distinction between

— 153 —

force and motion, for More identified the two (1974: 402–5). Because of his impending departure for Sweden at the time when More's third letter arrived in the Netherlands in August 1649, Descartes never found time to deal with the scientific issues that More raised in it, and extended in his fourth letter (23 October 1649) which never reached Descartes.

The matters for explanation in natural philosophy on which More requested further elucidation were many and difficult. For example, concerning Descartes's theory of the celestial vortices, More asks 'Why are not your vortices in the form of columns or cylinders rather than ellipses, since any point of the axis of a vortex is as it were a centre from which the celestial matter recedes with, so far as I can see, a wholly constant impetus?' (1974: 389). That is, the axis of rotation of a sphere must be a line, not a point. Then More puts astronomical difficulties, such as

> Who causes all the planets not to revolve in one plane (the plane of the ecliptic)? And also sunspots, not to be in planes at least parallel to the ecliptic? And the moon itself, neither in the plane of the Earth's equator nor in a plane parallel to this? Since these bodies are directed by no internal force but are merely borne around by an external impulse? (1974: 386)

Other questions relate to the specific 'mechanical models' devised by Descartes in order to explain optics and magnetism. For example, it seemed to More that because the eye can perceive different colours in close juxtaposition, the light-particles reaching it must be affected with multiple rotary motions at the same time, which is nonsensical; for Descartes specifies a particular rotary motion (or spin) associated with each colour (1974: 390). At this point, as the correspondence ceased (owing to Descartes's involvement with Queen Christina, then his premature death), it is evident that More was capable of raising difficulties about the Cartesian mechanical philosophy within its own terms, that is, of questioning its consistency and inherent plausibility. It may be doubted whether Descartes would have had better success in satisfying his critic at this level, than he had had with his metaphysics.

The cessation of Descartes's letters came as a great disappointment to Henry More; his design, he wrote to Hartlib, had been to gain a thorough understanding of Cartesian philosophy in order to become 'a safe and able judge of it'. 'Which I am the more eager of, because I have never mett with any naturall philosophy yett, that seemed to me worthy of so narrow a search, as this is'. And then More adds, strangely, that this philosophy would 'at the very first seizure upon it, ... fall a pieces like a rotten duck in a man's hand, or vanish like a mist, and leave nothing but a few empty termes, or Logicall notions, insted of substantiall truth' (1974: 633). Which may indicate that More was not so far from Christiaan Huygens's later description of Descartes's physics as a fine piece of scientific fiction.

It may seem that More's critique of Descartes's philosophy is unsystematic and lacks focus. More attacks at many points but without an obvious strategy, save that he starts with the *Principia philosophiae* and its treatment of motion, the vacuum, etc., then goes on to *Dioptrique* and *Météores*. The great issues of God, spirit, matter and motion are (one feels) important to More, who writes from the strength of his own opposite convictions; but the detailed questions about the various kinds of particle imagined by Descartes are of a different order, academic exercises that cannot have concerned him deeply. There are, however, two profound differences that separate the two philosophers. The first is not unlike that which divided Newton from Leibniz: it is a question of the ontology of spirit. Are God, the Spirit of Nature and individual spirits the sole sources of activity in the universe, rather than that everlasting motion which Descartes supposed to inhere in matter? The second difference is methodological: are metaphysics and natural philosophy closely coupled together – as Descartes would have it, the latter logically dependent on the former – or is it (in the limit) possible to match any natural philosophy (or physical model of the universe) with any type of metaphysics? Many subsequent philosophers have agreed with Descartes that a particular set of metaphysical assumptions entails a certain physical world-view; or conversely, starting from a given body of scientific knowledge only a limited number of metaphysical assumptions can be framed that are consistent with it. More continually pressed upon Descartes a different opinion and guided his own conduct and teaching accordingly: that a sound system of natural philosophy can be related to more than one metaphysics. In this case clearly the soundness of the system has to be tested independently of metaphysics, for example, by its inner mathematical coherence and its conformity with the findings of observation and experiment. It cannot be established, as it is in Descartes's philosophy, by an *a posteriori* agreement with a metaphysics that has been settled *a priori*. According to More, given that the philosopher believes (for whatever reasons) in the soundness of a particular body of natural philosophy, then he is entitled to match this with any set of metaphysical assumptions that he happens to prefer. In this way More anticipates loosely the attitude of some scientists in later times who have regarded metaphysics as, so to speak, a personal variable.

It was the central tenet of More's thought, life-long, that there is a subservient truth in the explanatory structure of atomism just as there is a higher truth in the creative, directive and energetic power of God and the spiritual agencies he employs. More regarded Descartes's philosophy as the most promising expression of this subservient truth in his own lifetime – for in his eyes Gassendi and his followers offered nothing more than Epicureanism rehashed (Gabbey 1982: 232).[6] Therefore, while making it clear that this philosophy was no more than an excellent sketch that demanded extension and perfection before it could be definitively evaluated – and it is perfectly obvious what More would have wished to add to it – More would not lightly accede to others' criticism of Descartes, still less accept outright condemnation of his system. The public avowal of his commitment continued from the 1650s well into the next decade, most markedly (and ironically) in More's insistence that Cartesian philosophy is not conducive to atheism. Thus in the Preface to his *Philosophical Writings*: 'My main design in that Letter [*Epistola H. Mori ad V.C.*] was to clear *Cartesius* from that giddy and groundless suspicion of Atheism ... which I conceive I have done fully.' And in fact the *Epistola* contains five pages devoted to the exculpation of Descartes from this charge (1664a: 34–9).

While publicly engaged in thrashing the alchemist Thomas Vaughan (verbally) for his contempt towards Descartes, assuring him that 'There never was any thing proposed to the world in which there is more wary, subtill and close texture of reason, more coherent unity of all parts with themselves, or more happy uniformity of the whole with the phenomena of nature' (More 1650: 89, quoted in Burnham 1974: 42) than in the writings of the French philosopher, in private More was just as keen to defend him against any untoward objections raised by his 'heroine pupil', Anne Finch, though freely acknowledging his own difficulties in understanding or following Descartes at every point. More began to correspond with John Finch's sister in August 1649, having previously sent for her use a partial English translation of the *Principia philosophiae*, adding the third part of the same book in February 1650 (Nicolson 1930: 51–2). He advised her 'to habituate yourself compos'dly and steddily to think of any thing that you think worth the thinking of, and to drive it on to as clear and distinct apprehension as you can', without prejudice to her health by excessive study. When she responded to his invitation to 'send to him' should she meet with difficulties in her pursuit of philosophy, he rejoined with a warm defence of Descartes's ontological proof of the existence of God and assured her that Descartes had not assumed the untenable position that we cannot be sure whether we wake or dream.[8] After her marriage and an episode of ill-health, Lady Conway had, by May 1651, progressed as far as Part II of the *Principia*, enabling More to explain that he and Descartes in effect agreed in supposing the universe to be infinite, and to argue against Descartes's denial of the vacuum. More's position was that there can be distance in a vacuity, because distance is distinct both from space and from extension; space is penetrable but also it is 'immovable, and impassible. All the porters in London will not be able to carry one foot square of it from Cheepsyde to Charing Cross' (Gabbey 1977: 389). Accordingly, 'empty space' implies no contradiction. More agreed with Lady Conway that *infinite* and *indefinite* 'come much to one'. After this point their philosophical correspondence turned to other topics less directly concerned with Descartes, but it is evident that Lady Conway tended to find the Cartesian identification of extension with matter convincing, while on the other hand she had difficulty in accepting the contention that, for example, colours exist only in the mind. She inclined 'to the belief of colour to be a reall thing, and that it is something more than Des Cartes allowes it to be' (1977: 397). Unfortunately we lack More's handling of this issue.

Henry More's subsequent development as a philosopher was to a very great extent determined by his reading of Descartes and by the necessity to express his thoughts in the letters addressed to the French philosopher. As Alan Gabbey has said, more particularly, 'The criticisms that appear in More's writings from the *Antidote against Atheism* onward can be found, either fully fledged or in embryo, either explicitly or implicitly, in the correspondence of 1648–49' (Koyré 1957: 125; Mamiani 1979: 89; Gabbey 1982: 193). All except one, that is, the accusation that the materialist philosophy of Descartes excludes God and so encourages atheism – the very point upon which in his earlier years More had asserted Descartes's innocence. More brought forward this charge for the first time in his *Divine Dialogues* (1668) and was to reiterate it with greater vehemence in *Enchiridion*

Metaphysicum (1671). But acute readers had perceived much earlier that the *Epistola ad V.C.* (1662) indicated a new turn in More's thinking. His sense of the errors of Descartes and of the dangers flowing from them was becoming more pressing.

At least one of More's contemporaries, his friend John Worthington, appreciated that his retreat from Cartesianism left readers of More's works in a state of perplexity and ambiguity. If materialism alone were insufficient to explain the world, and if room had to be found in the world-view for spiritual entities, what kind of philosophy would result? After his expulsion from the Mastership of Jesus College, Cambridge, in 1660, Worthington passed the remainder of his life in country livings, one of which was Ingoldsby, to which he was presented by More; in return, he seems to have scrutinized More's writings of this period in proof, serving as a private critic of his matter and presentation. One of his most important recommendations to More is in a letter of 29 November 1667:

> I wish you would refresh and joy me with the news of your being about a Body of Natural Philosophy. I cannot but think you are about it, and if you will tell me so, I will keep it private – you have as highly commended Des Cartes, as is possible, and as knowing no better method of Philosophy, you recommend it effectually in some parts of your books, whereby you had so fired some to the study of it, that your letter to V.C. (which came long after) could not coole them, nor doth it yet: but they are enravisht with it, and derive from thence notions of ill consequence to religion ... And seeing they will never return to the old Philosophy, in fashion when we were young scholars, there will be no way to take them off from idolizing the French Philosophy, and hurting themselves and others by some principles there, but by putting into their hands another Body of Natural Philosophy, which is like to be the most effectual antidote. And to do this will be more easy to you, than any, because you have so fully consider'd it, and the chief materials of a new Phisiologie, you have more or less treated of in your other writings, the substance of which you may transfer into this book. (Christie 1886: 254)

Unknown to Worthington, More had, at the time of his letter, already been working on *Divine Dialogues ... touching the Attributes of God and his Providence in the World ... collected and compiled ... by Franciscus Palaeopolitanus* for some eighteen months. It is a wordy and turgid book, by no means what Worthington asked for, though sometimes judged the most readable that More wrote. Marjorie Nicolson devoted some enthusiastic paragraphs to it (1930: 273). Few are now stirred by More's apocalyptic rhetoric, conventional enough:

> The *blasphemous Horn with eyes*, the *Whore of Babylon* and the *Man of Sin*, methinks, are as reprochfull Titles as that of *Antichrist*: and if the Bishop of *Rome* could be proved to be any of these, especially that *Man of Sin*, it would be hard to fend off that other more ordinary imputation which they so much whinch at. (1668, II: 125)[9]

Or even by More's opposition of the spiritual to the mechanical:

> the *Primordialls* of the World are not *Mechanicall*, but *Spermaticall* or *Vital*; not made by rubbing and filing and turning and shaving, as in a Turner's or a Blacksmith's Shop, but from some universal Principle of inward Life and Motion containing in it the seminal forms of all things, which therefore the *Platonists* and *Pythagoreans* call the great λογος σπερματιτης of the World. (1668, I: 37)

Certainly the conceit is witty: the mechanical philosophy savours too much of the rude mechanicals, conveys little sense of the divine skill and grace. But the ideas and arguments of this long book are simply picturesque restatements of what More had said before, sometimes in the same words. Once again he goes over his problems in pneumatics and hydrostatics: air and water can have no weight and exert no pressure in their proper places, for if they did the creatures immersed in them would be crushed to death. Philotheus asks: 'Must not then some divine Principle be at the bottom, that thus cancels the Mechanicall Laws for the common good?' And Hylobares replies: 'It should seem so: and that the Motion of Matter is not guided by Matter, but by something else' (1668, I: 34).[10] Once again More is puzzled to know why a light, thin board nearly fitting the interior of a cylinder full of water floats to the surface. It is evident that More is willing to argue against Descartes, as against Boyle and Hooke, that his philosophy 'fails so palpably even in the general strokes of Nature, of giving any such necessary Mechanicall Reasons of her Phaenomena' (1668: 39).

Perhaps more interesting than the text is the Preface to the first volume, where the Publisher (surely drawing words from More's brimming well) doubts whether the author has overshot himself in anything in the book

> unless in his over-plain and open opposing that so-much-admired Philosopher Renatus Des-Cartes, on whom persons well versed in Philosophicall Speculations have bestowed so high Encomiums, especially a Writer of our own, who, besides the many Commendations he up and down in his Writings adorns him with, compares him (in his Appendix to the Defence of his Philosophicall Cabbala) to Bezaliel and Aholiab, as if he were inspired from above with a Wit so curiously Mechanicall, as to frame so consistent a Contexture of Mechanicall Philosophy as he did.

However, the learned Publisher is able to cover More's retreat by drawing attention to the record of his reservations about Cartesianism, particularly as expressed in the *Epistola ad V.C.* 'where he makes it his business to apologize for him [Descartes], and to extoll him and magnifie him to the skies' (1668: Preface) but had also written that it would be a vile and abject idolizing of matter to pretend that all the phaenomena of the universe arise out of it by mere mechanical motion.

Descartes's assimilation to the grand 'Cabbalisticall' tradition as understood by More, formerly counting in his favour, is now turned against him. The chief merit of his philosophy is

> that he has interwoven into it that Noble System of the World according to the Tradition of Pythagoras and his followers, or, if you will, of the most ancient Cabbala of Moses. But the rest of his Philosophy is rather *pretty* than *great*, and in that sense that he drives at, of pure Mechanism, enormously and ridiculously false. (1668: Preface)

So, says More, where Descartes is sound and true he has borrowed from the Mosaic legacy; where he has introduced his own principles:

> there is nothing more estranged from the *Genius* of the Scripture and the services of Theologie than they.
> Let the zealous *Cartesian* read the whole 144 Psalm, and tune it in this point, if he can, to his Master's Philosophy. (1668: Preface)

All in all, More concludes, 'Religion can suffer nothing by the lessening of the Repute of *Cartesianism*, the Notions that are peculiar thereto having so little tendency to that service' (1668: Preface). More's attachment to ghostly and miraculous visions reinforces the same lesson; with the aid of irony: 'The Apparitions of Horsemen and Armies encountring one another in the Air, 2.*Macch.*5, let him consider how illustrable that passage is from the last Section of the 7. Chapter of *Des-Cartes* his *Meteors*, and from the Conclusion of that whole Treatise' (1668: Preface).

The *odium theologicum* thus figuring in the *Divine Dialogues* was given even stronger expression in *Enchiridion Metaphysicum* (1671), on which More must have set to work at once. It seems that More's hand could hardly work fast enough in pulling down what he had formerly built up:

> No greater wound or injury can be inflicted upon those elements which are most essential to religion, than by the possible presumption of the resolution of all phenomena into purely mechanical causes, not excepting the bodies of plants and animals. Which is the Cartesian hypothesis ...
>
> Truth to tell, in the state things now are, if the Cartesian philosophy in both its metaphysical and its physical aspects should stand firm, I shudder to think of the greatness of the inclination, and of the danger, for the souls of men to fall into atheism. (1671: Preface)

Certainly in the depths of his invective More did not accuse Descartes personally of being an atheist, or of seeking to promote atheism, but he does again and again assert (as here) that Cartesianism leads men into atheism. Without intending evil, Descartes has certainly caused it, not least by attracting such followers as Spinoza (p. 193). Descartes could hardly have done more damage to religion if his intentions *had* been evil.

Such scholars as Alan Gabbey and John Rogers have recently pointed out, rightly, that as More's dissent from Descartes was constant in character throughout his life, though increasing in vehemence, and as his adhesion to the philosophy of a 'Divine Spirit and Life' which in his view far excelled Descartes's as a 'true Ornament of the Mind' was unflagging, it is needless to seek for a violent turnabout in More's thought which sent Descartes from the zenith to the nadir of his esteem (Gabbey 1982: 204; More's words from a pamphlet against Thomas Vaughan, 1651). What we see, and need to account for, is a change of balance: Descartes, having been a safe (though imperfect) guide, became from about 1660 a philosopher whom the Christian should follow only with grave caution. Since More was never formally an expositor of Descartes's philosophy or his intellectual disciple, the exaggerated claim for him as the first English Cartesian has created mis-apprehension. Any comparison of Henry More with, say, Jacques Rohault would be absurd. As Gabbey says, More adopted, adapted, used – one is tempted to say exploited – Descartes for his own ends. More's objectives in philosophy were never those of his greater French contemporary, but More realized that the two of them could travel together for part of the journey. More always believed, with reason, that his concern for religion was far more urgent than that of Descartes. And (let us note once more) Henry More's devotion to Platonism and the perennial philosophy were constant features of his intellect that Descartes did not share. Thus More's Cartesianism was highly flexible. If it could be employed as a weapon against irreligion, More was prepared to adopt it; if it seemed that the same weapon might cut more sharply against religion, then More would fight against it.

This, roughly speaking, was also the opinion of Richard Ward in his not insensitive comments on More's change of allegiance; the more 'our Authour' considered the mechanical philosophy (he wrote in the unprinted section of his biography),

ye more he took it to be of dangerous consequence, as well as absolutely impossible; and hath given ... Clouds of Reason for it; and this out of ye truest Honour and Loyalty to ye Cause of God, and Interests of Religion; and because of ye very ill purposes yt he saw this French philosophy actually both abused, and abusable to; and to cut off for ye future ye occasions of it. And this with a serious particular yt will be hinted, is that wch made him so zealous afterwards against ye Cartesian Philosophy; after all his high elogisms of it and of ye Authour; in wch he owns, yt through his extreem Admiration of at first, and deference to yt great Man, there was an excess; but yt it only proceeded from a most prompt Mind, and ready disposition, to receive and applaud wtever seemed eximious, let it come from wt Hand or Quarter it will; not from any turgid Emulation or Pride. Des-Cartes his Metaphysicks I could never perceive yt he much admird; but his Physicks he did exceedingly: and thinks it pardonable if in his Youth, or younger time, and under ye noble surprise and Ingenuity of ye Thing, he saw not into all its Flaws or Consequences at once, but extolld it after a higher and more unlimited Manner then otherwise was at all fit, or he could afterwards doe. (MS, Christ's College, Cambridge)

It is useful in discussing the chequered history of More's Cartesianism to distinguish, as Ward did, More's treatment of Descartes's metaphysics and theology from his attitude to Descartes's natural philosophy. The former was more constant though not less critical. As his own ideas developed, More did not find fresh occasions to dissent from Cartesian metaphysics nor were his dissensions ever so serious as to shade his initial high esteem of the Frenchman. Others, better qualified to discuss such themes than I am, may take a different view of this aspect of the matter; for myself, I cannot believe that the metaphysical points at issue, though they bulk large in More's letters of 1648–9, would have led him to the warmth of his later denunciation of Cartesianism. It was with regard to Descartes's natural philosophy, that is the mechanical philosophy of Nature, that More's change of opinion was profound. Not that More ever rejected his own 'Platonizing' version of atomism, always opposed to that of the classical tradition from Democritos to Epicuros, which was to become equally hostile to the matter-theory of Descartes. To More's mind, atoms were simply building-blocks; using their variations in size and shape, the philosopher could usefully construct hypotheses about the fabric of the universe. (More would surely have approved of Daltonian chemical atomism, which does just this.) But further, in More's view, whatever might be said about the cohesion of atoms and their separation, or their movement of any kind, was not to be said of the atoms themselves but of whatever cause brought these changes about. Matter itself was totally inert, and could only submit to motion, not generate it.[11] In More's game of billiards, movement is in the cue, not the ball. As the analogy indicates, the real difficulties for More concern *inertia* and *quantity of motion* – the fundamental concepts of the science of mechanics. Is that moving red ball ontologically distinct from that motionless white ball? Descartes and Huygens, the relativists, denied that it is; More and Newton, the absolutists, affirmed the proposition (and Newton called inertia a *vis insita*, an internal force). However this question was answered, it was certain that *two* balls moving relatively to each other were ontologically distinct from two balls at rest – a distinction measurable in terms of weight and speed (or mass and velocity), whose product came to be called the quantity of motion.

Because More does not present his own thoughts about the foundations of mechanics plainly, I shall venture upon a reconstruction of them. He might have objected to Descartes that to reify motion, to give it ontological reality and quantify it, is like talking of a quantity of taste or a quantity of fairness. All that we observe is a body, A, more or less large, shifting from position (1) to position (2), more or less quickly. We cannot infer from this the existence of an abstract entity 'motion', which Descartes then proceeds to cut up into parcels of different sizes, as though it were a weightless fluid, attaching each parcel to a smaller or larger bulk of matter. And because there is no such entity, it cannot be further shared out, or redivided, between the particles by the process of their mutual impact, as Descartes supposed. This is what he wrote to Descartes in his third letter (23 July 1649):

> Moreover, I consider the first principles so scrupulously that already a new difficulty about the nature of motion occurs to me. Since the motion of a body is a mode, like its shape, the arrangement of its parts etc., how can this mode be transferred from one body to another, rather than any other mode of bodies? And generally speaking my imagination cannot comprehend how anything that cannot be external to the subject (and this is true of all modes) can migrate to another subject. Hence I ask: when one body strikes another lesser one at rest and carries that along with itself, does the motionlessness of the resting body similarly migrate into the body bearing it along, as the motion of the latter migrates into the former? For it seems that rest is so lazy and inert, that movement would weary it; yet since it is as real as motion, reason dictates that it should migrate. (Descartes 1974: 382, my translation)

More's question, by which he thinks to embarrass Descartes, is not unsubtle; if rest and motion are equivalent states, as he claims, the former must be shared between the two bodies colliding, just as the motion is. But, in any event, Descartes's idea is false because the particle or body endowed with a quantity of motion, like a rider upon a horse, becomes self-propelling, and More believes (like Aristotle) that every moved thing has a mover, because inert matter cannot move itself. When the philosopher talks of motion, he should not talk of *moved things*, as Descartes does, but of *movers*, that is, spirits.

In More's eyes, Descartes's philosophy sinks even further into error and absurdity when he maintains that the totality of all the little parcels of motion attached to the totality of the particles of matter is constant within the universe. More believed, and Newton followed him in this, that motion can be locally destroyed and created anew. But of greater import is the conclusion from Descartes's position that just as each particle may be self-propelled, so the whole universe impels itself in all its phenomena. All are reduced to the motions of particles and each of these particulate motions is self-generating, or worse, may be said to be generated by matter. Such a universe is totally materialist, godless, and More holds that it is just as unthinkable as the universe of Epicuros, not only because it excludes the *active* power of God, but because it also denies the *directive* power of God. According to More, though God might have planned the arrangement of the particles at the creation of the universe and conferred appropriate motions upon them (this was Descartes's teaching) it is impossible that the mere mechanical exchange of motion among particles could produce regularity in phenomena. (More might have imagined a billiard table and a number of balls, all being perfectly frictionless and perfectly elastic; if one ball is struck by the cue the resultant motion – of one ball or many at different moments – will continue for ever. But the movements will be random, unpredictable.)[12] And More is right to the extent that the bulk of the natural philosophy in Descartes's *Principia* is devoted to the description of arbitrary and rather far-fetched mechanisms for ensuring that the partition of the motion in the universe among the particles does *not* take place in a random way, but only along particular pathways so as to produce regular patterns of phenomena – gravity, light, magnetism and so on. The More writing after 1668 no longer found Descartes's attempts to force order into randomness convincing: 'the unsuccessfulness of his Wit and Industry in the Mechanicall Philosophy has abundantly assured the sagacious, that the *Phaenomena* of the Universe must be entitled to an higher and more divine Principle then mere Matter and Mechanicall Motion' (1668: The Publisher to the Reader). And Descartes's failure was not due to his *a priorist* approach to natural philosophy, for the experimentalists Boyle and Hooke fared no better (chapter 9). All the various models of vortices and pressures, of springy particles in air and screwed particles in magnets are failures, says More, (1) because they commit the old pernicious error of attributing activity or motion to inert matter (how could inert substance be elastic?); (2) they ignore the necessity for the guidance of the Divine Hand and Providence; and (3) they simply do not pass the tests of experience.[13]

In both *The Immortality of the Soul* and the *Divine Dialogues* More approves the heuristic use of the mechanical philosophy – its evident limitations encourage belief in the Higher Power and the Spirit of Nature that he proclaimed (Gabbey 1982: 237). But his confidence in the wisdom of such a use seems to have lessened somewhat through the years as he realized that some readers of Descartes were so morally swayed by his arguments and models that they became atheists.

Why did More become so frantic – 'enthusiastic', John Beale had said – in his opposition to the mechanical philosophy? For we must recognize that in More's eyes the experimentalists Boyle and Hooke were as culpable as Descartes, or only slightly less culpable because they did not offer a full philosophical defence of their pernicious errors. All three were equally misguided in their denial of the obvious activity of Spirit in the world (though not malicious) but it would have been absurd in More to accuse Boyle of directly endangering religion. Therefore More was content to attack the Englishmen's phenomenalist exposition of mechanism at the same level. With this, More seems to have joined – as Cudworth did not – in the wave of academic reaction that swept over Europe in the 1660s. Whether More was caught up in a movement he did not help to originate, or whether he actually helped to start the movement of reaction, it would be hard to say.[14] In the Netherlands, the spirit of reaction inspired denunciations of the infiltration of Cartesian teaching into the universities and worked up tremendous ill-will against Spinoza. In England criticism of Hobbes and his few followers, and the anti-Royal Society polemic of Henry Stubbe, were its chief manifestations; only a few scholars began to look for enemies of tradition and atheists under every bed. More, to his credit, was never blatantly anti-scientific nor pro-Stubbe. But his letters to Lady Conway, as well as his publications, demonstrate his feeling that the Dutch academics had the right of it in making vigorous attacks upon 'atheistic' tendencies in philosophy.

Since More revised his judgement of the mechanical philosophy in general, and Cartesianism in particular, does this indicate some failure of comprehension on his own part during the 1640s and 1650s? I think it does. The mechanical philosophy did not change its character, nor did any development of its scope and application so modify it as to confront More with a different philosophical situation from that prevailing during the first ten or fifteen years of his 'Cartesianism'. Nor was there ever anything obscure about the notorious mechanism–atheism association in the case of Thomas Hobbes (who was never the major focus of More's attention). As regards mechanical hypotheses in natural philosophy, such as the 'spring and weight of the air', it may well be that More was simply ignorant of them (in the writings of Torricelli and others) until instructed by Boyle and Hooke. As for the Cartesian philosophy, we have to suppose either that the younger More failed to perceive the difficulties that (to his mature eyes) rendered it untenable, or that he perceived them as being too slight to impair the generally excellent and praiseworthy character of Descartes's writings. Since More's letters to him invalidate the first alternative (at least in part) we must conclude that More was prepared for a decade or more to set aside his differences from Descartes, for the sake of the increase in understanding provided by his philosophy. I think it was only slowly that More realized that (1) Descartes's idea of 'perpetual motion' in the universe was *intended* to make any appeal to 'natural spirits' (in More's sense) within philosophy redundant, and (2) some readers of Descartes – as well as Hobbists – agreed that spirits were redundant and God a retired Clockmaker. The notion that Descartes had improperly banished God from his own creation was not, I think, put forward by More in 1648–9; it was only in *Divine Dialogues* twenty years later that he began to write contemptuously of 'Nullibists' (those who admit the existence of a deity but can find no place in the universe for him) whose prince Descartes was. If this interpretation is correct, then it follows that More had entertained an imperfect conception of the revolution in scientific thinking in which Descartes was a prime mover; for he had failed to understand that, though that revolution was intended rather to strengthen than to weaken rational religion, it advanced a truly natural rather than transcendental philosophy. It could embrace vitalism, but not More's animism. In the sense of that *reason* upon which Cambridge Platonism had insisted so firmly, the new movement in natural philosophy had already chosen a position antithetic to that which pleased More's mind.

9

More and the Royal Society

THE ROYAL SOCIETY OF LONDON for Improving Natural Knowledge, to cite its later title, originated in a meeting of interested persons that took place at Gresham College, London, on 28 November 1660. Those present had just listened to a lecture by Christopher Wren, the Gresham Professor of Astronomy, after which they had withdrawn 'for mutual conversation, into Mr Rooke's apartment, where, amongst other matters discoursed of, something was offered about a design of founding a college for the promoting of physico-mathematical experimental learning', so that their 'mutual conversation' might be transformed into 'a more regular way of debating things with a view to the promoting of experimental philosophy' (Birch 1756, I: 3).[1]

This original group, each one of whom was to play an important part in the future Royal Society, appointed John Wilkins (1614–72) to be its Chairman *pro tem.* and proceeded to draw up a list of the names of others who should be invited to join the nascent college (or, rather, Society). There were no Cambridge Platonists on this list, but the name of Henry More's younger friends, John Finch and Thomas Baines, were included and both became in due course Original Fellows of the Royal Society. Wilkins, at this time Vicar of St Lawrence Jewry (a living in which he was followed by Benjamin Whichcote on Wilkins's being promoted to the See of Chester in 1668), stood high in the Royal favour as did another leading figure in the group, Sir Robert Moray; these especially won royal approval and name. Wilkins had presumably become personally acquainted with both Cudworth and More during his brief Mastership of Trinity College during the previous year, and must also have read More's writings. However, it was not until 4 June 1662, when Wilkins was again President *pro tem.* of the meeting, that he proposed the two Christ's men as members of the Royal Society. They were duly elected (together with Isaac Barrow, Gresham Professor of Geometry in succession to Laurence Rooke) on 17 September. As 'country members', whose frequent presence at meetings in London was not to be expected, both were excused the payment of the subscription. Nothing indicates that either Cudworth or More ever attended a meeting of the Royal Society at this time or indeed were informed of their election. Accordingly, when Charles II granted a second Charter formally incorporating the Royal Society on 10 May 1663,[2] with a list of those who came to be called Original Fellows, neither Cudworth nor More figured there (nor were they unique in being so omitted). The latter was proposed for membership a second time by John Wilkins over a year later (25 May 1664) and was admitted to the Fellowship one week later.

After this formal act of attendance, we learn from Pepys's *Diary* of More's presence also at one of the great gala days in the Society's early history – its entertainment of the bluestocking Duchess of Newcastle, on 30 May 1667:

> Several fine experiments were shown her of Colours, Loadstones, Microscope, and of liquors: among others, of one that did while she was there turn a piece of roasted mutton into pure blood – which was very rare – here was Mr. Moore of Cambridge, whom I had not seen before, and I was glad to see him – (Birch 1756, I: 427; 432; 435. See also Hunter 1982; Pepys 1974, VIII: 243)

One later entry in the Royal Society's records refers to More by name. On 25 February 1685 one of the Secretaries read a letter to himself from Isaac Newton, who was now (as he wrote) deeply engaged in the composition of his *Philosophiae naturalis Principia mathematica* (Birch 1757, IV: 370; Newton 1959, II: 415). (This Secretary, Francis Aston, a contemporary and friend of Newton's at Cambridge, had played a part in bringing this work before the Royal Society.) In his letter to Aston, Newton told how he had engaged Dr More to be a member of a 'Philosophick Meeting' to be established at Cambridge. This is one of the rare evidences of personal contact between these two men. More was about 70 years old, strong neither in health nor in body, and Newton can hardly have expected him to be an active promoter of experimental investigation in Cambridge. The embryonic scheme collapsed because, as Newton went on to say, there was a 'want of persons willing to try experiments, he whom we chiefly relied on, refusing to concern himself in that kind'. Newton was just as reluctant to 'engage the loss of his own time in those things' by undertaking communal experiments. Who the person was remains a mystery: certainly not More. Yet there are other references to an air-pump at Christ's College, and More certainly possessed scientific apparatus, so it may be that Newton and his friends had had their eye on some other member of that College such as Thomas Standish or Thomas Lovett.

Henry More's allusions to the Royal Society are infrequent. He did, however, draw a distinction between the *mechanical* philosophy – which, from 1670 onwards, he increasingly distrusted as conducive to atheism – and *experimental* philosophy which could properly be pursued without metaphysical or theological emperilment. So in the Preface to *Enchiridion Metaphysicum* (1671), after sketching the iniquity of Cartesian materialism, he continues:

> I mean all these points to be so understood, that they in no wise depreciate the worth and value of the Experimental Philosophy, which many improperly confuse with the Mechanical; such philosophy namely as the celebrated Royal Society of London professes, in which kind it has produced many and notable exemplars of its art and industry, not only of benefit to the conveniences of life but (as this *Enchiridion* makes evident) of especial usefulness in the discovering of philosophical truths and particularly those that are sublime and truly metaphysical.

This reads well as a tribute, and may represent some polite concession in More: as we shall see, he was far from being in complete agreement with the founders of the Royal Society touching the importance and philosophical intepretation of experiments.

Newton's letter serves as a reminder that the man whom Newton had invited to dignify a philosophical club had been, in his time, a philosopher noted for his acceptance of the latest discoveries about the natural world. More's early publications show him occupying much the same role of vernacular publicist as John Wilkins at the same epoch, and there is no reason to suppose More's attainments in mathematics and natural philosophy much below those of the much-praised Wilkins. The latter developed the 'science-fiction' mode begun by Kepler: More preferred the vehicle of metaphysical didactic verse. The four poems of 1642 not only proclaim an exultant Platonism (a motto from Hermes Trismegistus, no doubt taken from Marsilio Ficino, is prefixed to each) but a clear vindication of Copernicus against Ptolemy. As the following lines – published in the year of Galileo's death – reveal, More welcomed the Copernican universe because it was idealistic and Pythagorean:

> I have the barking of cold sense confuted ...
> To show that Pythagore's position's right.
> Copernicks, or whosoever dogma't hight.
> (1647: 'Psychathanasia', Book III, Canto 3, Stanza 44)[3]

He was thoroughly familiar with the technical details of heliocentric astronomy, and with the natural-philosophical arguments employed in its defence by Galileo and others. Thus More disposes of a well-known 'mechanical' objection to the motion of the Earth cited by the anti-Copernican Tycho Brahe:

> An arrow shot into the empty aire,
> Which straight returning to the bowman's foot,
> The earth's stability must for ever clear.
> Thus these bad archers do at random shoot,
> Whose easie errour I do thus confute.
> The arrow hath one spirit with this sphere,
> Forc'd upwards turns with it, mov'd by the root
> Of naturall motion. So when back't doth bear
> Itself, still Eastward turns with motion circular.
> (1647: 'Psychathanasia, Book III, Canto 3, Stanza 37)

A late allusion – one of few – to More's teaching of astronomy occurs in a letter to Lady Conway of 1674:

> I wish your Intermissions [of her illness] would permitt your Ladiship to
> recalle to mind the acquaintance you had with Copernicus his Hypothesis,
> the Motion of the Earth, that you might examine Mr Edward Rawdon [her
> nephew, More's pupil] in it, when he wayts on your Ladiship. I used that as
> an argument to him to be well skill'd in the Hypothesis. (Nicholson 1930: 388,
> 12/4/1674)

The nephew of this extraordinary aunt could have found the outline at least
in More's philosophical verse. Four years after the publication of *Psychodia
Platonica*, after reading Descartes's *Principia philosophiae*, More in *Democritus
Platonissans, or An Essay upon the Infinity of Worlds out of Platonick Principles*
stretched his poetic vision to a far bolder characterization of a universe
infinite in space and time:

> Wherefore at once from all eternitie
> The infinite number of these Worlds He made,
> And will conserve to all infinitie,
> And still drive on their ever-moving trade,
> And steddy hold what ever must be staid;
> Ne must one mite be minish'd of the summe,
> Ne must the smallest atom ever fade,
> But still remain though it may change its room;
> This truth abideth strong from everlasting doom.
>
> (1646: stanza 70)

More spurns Descartes's prevarication with the word *indefinite* in place of
infinite: there is space wherever God is, as he was to repeat many times
thereafter (1646: Preface).

Not only is the universe infinite, it is also (as the astronomers teach) full
of other worlds than ours, with the whole starry creation arranged in due
order though lacking obvious symmetry: the stars are sown in the sky like
a 'peck of peasen rudely poured out / On plaister flore ... [in] harsh
disorder'd order':

> But though these lights do seem so rudely thrown
> And scattered throughout the spacious skie,
> Yet each most seemly sits in his own Throne
> In distance due and comely Majesty;

And round their lordly seats their servants hie
Keeping a well-proportioned space
One from another, doing chearfully
Their dayly task. No blemmish may deface
The worlds in severall deckt with all art and grace.

(1646: stanzas 54; 55)

Why the distances between the stars (or suns as More also calls them) are so 'perplexed', More says, 'must remain as yet unknown'. He seems not to have known of Kepler's Laws; epicycles he treats with derision.

All this, of course, is so much evidence of God's design and wisdom in the creation:

How long would God be forming of a flie?
Or the small wandring moats that play i'th' Sun?
Least moment well will serve none can denie.
And cannot He make all the World so soon?
From each atom of the matter wide
The total Deitie doth entirely won,
His infinite presence doth therein reside,
And in this presence infinite powers do ever abide.

(1646: stanza 69)

At one point, in pursuit of this pattern of analogy, More makes a curious scientific prophecy which the telescope was to verify in his own lifetime:

Besides each greater planet th'attendance finds
Of lesser. Our *Earth*'s handmaid is the Moon,
Which to her darkned side right duly shines,
And *Jove* hath foure, as hath been said aboven,
And Saturn more than four if the plain truth were known.

(1646: stanza 30)

The *Psychodia Platonica* and its successor show few if any signs of the authority of Descartes, rather the thought behind the verses seems to be that of Galileo, but More has always been credited with having the lion's share in the introduction of Cartesian natural philosophy into Cambridge, and indeed into England (see above). This alone would have merited his

— 173 —

election to the Fellowship of the Royal Society. Nothing survives to show what reading More gave to his pupils during the 1650s and 1660s, but once again his letters to Lady Conway furnish some account of his method during the penultimate decade of his life: he writes to her of her nephew: 'Mr Edward [Rawdon] comes diligently to his Cartesian lecture, we are just now gott through the 3 first parts of his *Principia*' (Nicolson 1930: 393, 17/9/1674). It seems evident that not only were such works as the *Discourse on Method* and the *Meditations* not treated by More, but that the emphasis was strongly on Cartesian natural rather than moral philosophy. So, in the following month, 'I have begun to read to him Des Cartes Dioptricks.' And having got them through that in some six weeks, More writes that he has made his pupil 'understand them from the beginning to the end, the Machine for making glasses not excepted. I was thinking to the Mathematical Demonstration' of the reflection of an image to a focus of the parabola, but has chosen instead to embark upon the *Meteors* (1930: 395, 19/10/1674; 397, 3/12/1674).[4]

Caetera desunt. The young man went off to Paris. But it is evident that still, a decade after his renunciation of Cartesian philosophy, More believed it to provide the best scientific grounding for the young. His attachment to Cartesianism served as a potential bond with many other Fellows of the Royal Society; his ability to take pupils through the *Dioptrics* and *Meteors* demonstrates no mean ability in geometry.

Although he was to prove a wholly inactive Fellow, the Royal Society played an indirect part in More's life because of his connection with those involved in its origin and evolution. Samuel Hartlib, the German-born intellectual organizer and intelligencer, whose influence may be discerned within the various currents that brought the Royal Society into being, was in correspondence with Henry More from at least the end of 1648, as well as with More's close friend John Worthington, the latter exchanges continuing until Hartlib's death in 1662. It was Hartlib who transmitted the letters exchanged between Moore and Descartes.[5] He was a great admirer of More's writings, especially of *The Immortality of the Soul*, which he commended to all his circle; and took steps to introduce its members to More on his later visits to London. So More in 1659 met the mathematician John Pell and Sir William Brereton, both future Fellows, though he possibly did not meet Hartlib himself until May or June 1661 (Crossley 1847: 140–2, 20/7/1659; 323, and many other references). Moore had never been strongly drawn to Hartlib's schemes, and felt some jealousy for academic leadership in intellectual matters. He must be presumed to have been acquainted with the other Cambridge men who were early Fellows, such as the naturalist John Ray of Trinity College. The most scientifically energetic, the most pious, and the best-known of all those with whom More was acquainted was Robert Boyle (1627–91), who was just entering upon his mature period of achievement when the Royal Society was founded. As to religion, there could be little weighty difference between More and the author of *The Christian Virtuoso* and *Theodora*; on the other hand, as we shall see later, the two men could not see eye-to-eye on the role of spirits, or occult causes, or the Divine Hand, in the explanation of natural phenomena. To Boyle, as to most Fellows of the Royal Society then, natural philosophy ended where the discussion of spiritual agencies and divine providence began, though all agreed with More's argument that the perfection of Nature demonstrated the existence of God. What few would do was to go along with More in his claim that phenomena daily revealed spiritual, rather than mechanical, agencies at work. At this level, accordingly, More was out of step with the Society of which Boyle was the glory and the epigone, and one might imagine that More would have been highly critical of the sceptical, even materialist, mentality in which the Society, whose motto was *Nullius in Verba*, carried on its enterprise. However, the Society's scepticism, its departure from tradition, and its materialism were more rhetorical than real; it provided no refuge for Thomas Hobbes. More did not set himself up in opposition to the Society's aims and methods and he was a close ally of one of the Society's chief defenders and apologists, Joseph Glanvill (1636–80), who was also no less convinced of the reality and effect of the spiritual world than More himself.

In the Epistle Dedicatory to his first book, *The Vanity of Dogmatizing* (1661) Glanvill relates how he had composed a work on the soul's immortality before the Restoration of the monarchy, which he had been content to suppress because of the 'maturer undertakings of the accomplisht Dr H. More'. This rather distant phrase does not suggest any close link between the two men at that time; as an Oxonian, recently ordained and barely out in the world, Glanvill would have had little opportunity yet to encounter More.

The theme of *The Vanity of Dogmatizing* may be defined in six of Glanvill's own words, echoing Nicolas of Cusa: 'The *knowledge* I teach, is *ignorance*.'

After giving instances of *what* we do not know in natural philosophy (how the soul moves the body, how gravity makes bodies fall, and so forth) Glanvill provides physical and psychological *reasons* for our not knowing, or perhaps lacking the capacity to know. Mathematical logic itself seems full of irresolvable puzzles (he instances Aristotle's paradox of the wheel). Therefore it is folly to dogmatize, to declare certain what must remain dubious. It soon becomes clear, however, that this is a folly to which the ancients (the Aristotelian philosophers) have been far more prone than the moderns (Copernicus, Galileo, Gassendi, Descartes, More himself). True, these have no more right than Aristotle to declare that they *know* with absolute certainty how the world is made, but they seem to have a more probable idea of it than the ancients did, especially 'the Grand Secretary of Nature, the miraculous *Des-Cartes* [who has] here infinitely out-done all the Philosophers [that] went before him, in giving a particular and *Analytical* account of the *Universal* Fabrick' (Glanvill 1661: 211). Yet Descartes (as Glanvill too disingenuously continues) 'intends his Principles but for *Hypotheses*, and never pretends that things are really or necessarily, as he hath supposed them'. Glanvill was neither mathematician nor experimentalist, nor perhaps a critical reader of Descartes, but he was emphatically attaching himself to the modern school. Like others among its defenders, he was unwittingly unfortunate in selecting for his argument such assured matters-of-fact as the Powder of Sympathy, 'put out of doubt by the Noble Sir K. Digby' because of the unexceptionable proofs given in his ingenious discourse concerning it; or the 'wonderful *signatures* in the *Foetus* caus'd by the Imagination of the Mother'; or the 'artificial *resurrection* of *Plants* from their *ashes*, which *Chymists* are so well acquainted with'. (1661: 47; 199; 207).[6]

Henry More is mentioned in a number of places, as when the tale of the mind-forcing scholar-gypsy is marginally justified by allusion to 'the *Hypothesis* of a *Mundane* Soul, lately reviv'd by that incomparable *Platonist* and *Cartesian*, Dr H. More', which indeed might be invoked to account for any strange 'sympathies'. Again, the delusions of imagination causing credulity to be cried up for faith have been exposed by 'the excellent Dr More [who] hath followed the *Enthusiastick effects* to their proper *Origine* ... His Discourse of *Enthusiasme* compleatly makes good the title [*Enthusiasmus Triumphatus*, 1656], and 'tis as well a *Victory* as a *Triumph*.' He applauds More's refutation of Hobbes's materialist theory of memory. But Glanvill does not, curiously enough, eagerly embrace the 'ingenious hypothesis of the most excellent *Cantabrigian* Philosopher of the Soul's being an *extended penetrable* substance', or at least choose to adopt this hypothesis in order to explain the soul's action upon the body; and he is careful elsewhere to endorse the majority opinion that 'the *Soul* and *Angels* are devoid of *quantitative dimensions*' and that such spiritual entities cannot properly be said to occupy a place, any more than thought or virtue can (1661: 22; 38; 100; 199). Now the real presence of spirit in matter – though not of course under any restriction as to form and volume – is a central tenet of More's cosmic system, and he specially introduced the idea that spirit can expand and contract in its extension (unlike matter), and be penetrated by matter within its extension, in order to make this coterminous coexistence of spirit and matter more comprehensible. Thus Glanvill was far from being a literally faithful disciple of Henry More. Too much, perhaps, was the influence in spiritual matters the other way.

Contact between the two men seems to have been literary rather than personal. Glanvill was a West of England man, associated with Exeter and Lincoln Colleges at Oxford and possibly with Wilkins and other 'new philosophers' there. Although a major apologist, self-appointed, for the Royal Society, Glanvill was little concerned in its affairs; his principal Church preferment was at Bath, not yet a fashionable resort. More took a slightly distant but approving view of him. 'I am very well assured of his vertu and probity, and that he is basely defamed by some former Enemyes of his,' he told Lady Conway (Nicolson 1930: 327–8, 14/3/1671). Later More was to collaborate with Glanvill in maintaining the reality of spiritualist phenomena, with Glanvill as the more vociferous partner, but the first consequence of Glanvill's attachment of himself to More was More's involvement in Henry Stubbe's attack upon Glanvill's partiality for the Royal Society and its objectives as displayed in *Plus Ultra* (1668). Stubbe had set himself up in medical practice in Warwick, not far from Ragley, where he and More had met at the time of Greatrake's abortive visit in 1666. The contemporary chronicler of Oxford University, Anthony Wood, attributed to him a 'hott and restless head (his hair being carrot-colour'd)' and an absence of fixed principles (1817, III: 1071). Everything that passed through his head came out of his mouth; in 1671 Glanvill told More 'how grossly and vehemently Stubbes rayled at me [More] in the coffy houses in Oxford [where] Stubbes [was] so generally lookt upon ... for a madman that he was asham'd of his

Adversary' (Nicolson 1930: 327, 14/3/1671). The ornithologist Francis Willughby – whose home was in Warwickshire also – sent More a copy of what Stubbe had written about him in *Legends no Histories*, as did Glanvill too, and therefore More 'was feigne to write a pretty long letter to J. Glanvill, which I beleeve he will publish. I think Dr Stubbs [sic] because he phancys me pious thinks I am a fool. When our Saviour Christ bids us be as wise as serpents as well as innocent as doves.' In fact More thought Stubbe 'is either mad or worse to stick [? pick] me up as a clod in the field to pelt the Royal Society with and putt me against them that from the beginning have been civill to me' (1930: 303, 6/8/1670).[7]

In the course of his polemic against Glanvill, Stubbe had accused him of favouring mechanism, citing More on the other side, as agreeing with Stubbe that the mechanical philosophy was pestilential and pernicious. Because the Royal Society had embraced this philosophy More, Stubbe alleged, had withdrawn from it. At Glanvill's request, More wrote him a letter expressing his surprise that Stubbe should know so much more of the matter than he did himself, since he was not aware of resigning from the Society. The preface to *Enchiridion Metaphysicum*, More continued, had made plain his esteem for the Royal Society's experimental philosophy: 'And [I] do particularly note how serviceable their *Natural Experiments* in *matter* are to the clear knowledge and Demonstration of the *Existence* of *immaterial Beings*: So far are they from tending to *Atheism*.'[8] It was a mistake on Stubbe's part to suppose that the mechanical philosophy opposed by More had been adopted by the Royal Society. For the philosophy that the Society was aiming to build, to be raised out of faithful and skilful experiments, is a more perfect philosophy which would rout both atheism and mechanism at once. Again, Stubbe was wrong to think that the mechanical philosophy consisted in the attribution of all 'the *Vicissitudes* of *corporeal* Nature' to the unions, dissolutions and motions of material particles; for this was but a 'part of the old *Pythagorick* or *Mosaic Philosophy*'. The error of the mechanists lay in thinking of the activity of the particles as self-originating and self-sufficient. The irrepressible Stubbe returned again with *An Answer to the Letter of Dr. Henry More* and *A Censure upon Certain Passages in the History of the Royal Society* (1671), but More thought himself fully excused from carrying any further a dispute with 'such an unworthy person' (1930: 327).

The rightness or wrongness of the Royal Society's methods and activities were not at all More's concern; clearly, he favoured its modernism and had been busily engaged in eroding the traditional philosophical teaching of the Universities long before the Society was founded, though this does not mean that he did not love their privileges (Syfret 1950: 34n). It is unlikely too that More had anything of Stubbe's supposed conviction that the Society was placing the Anglican Church in jeopardy – a charge that had some influence in Oxford but not, apparently, in Cambridge, which was at least as much opposed to 'atheism'. Even after his revulsion from Cartesianism, More was not at all inclined (as we have seen) to tar all neoterics in philosophy with the brush of materialism, and he must have known that many Fellows shared his view that the cosmic order was spiritual as well as material – not least two Fellows of Trinity College, Thomas Gale and Isaac Newton, besides his friend Ralph Cudworth.

It has been argued that More, an idealist and a Cartesian, lacked sympathy with the Baconian tradition in seventeenth-century England which a number of historians have identified as providing the ideology characteristic of the Royal Society (see for example Purver 1967; Webster 1975). To enter into detail on this large question would be out of place here: it is sufficient to draw attention to the existence of those elements in the conception and evolution of the Royal Society which were non-Baconian in origin though far (it may be) from being anti-Baconian; the elements concerned with the mathematical and medical sciences, for example, the traditions of Thomas Harriot and William Harvey rather than of Bacon, not to say Platonism. Whether this non-Baconian element in the Royal Society was large and influential or otherwise is an irrelevant issue here; the significant point is that there was *some* element in the Royal Society with which More could wholeheartedly sympathize. Moreover, those historians who have argued persuasively for the Baconianism of the early Royal Society have tended to forget that in addition to aspirations, ideas of method, logical principles and social objectives commonly shared by members of the Society (though never all), the inspiration and origin of which may be sought in Bacon's writings, there was a world-view equally commonly shared among them: a world-view that was modernist and mechanical, that embraced heliocentricity and the circulation of the blood, that accepted the new instrumental revelations about the nature of the universe, the new optics and the new pneumatics. Now it is undeniable that if we may speak of a new world-view widely held by the informed members of the Royal Society, it had not been shaped by Bacon who had rejected virtually every scientific innovation of his own lifetime. This world-view derived from such figures as Harriot and Harvey in England, more importantly Copernicus, Galileo and Descartes, all innovators with whom (as we have seen) Henry More was in confessed sympathy. No one would claim that the spiritualist features in More's personal world-view, increasingly emphasized in the years following his election to the Royal Society, were widely accepted by other Fellows, but the world-view that he had propounded before his election, during twenty years, which was Copernican, Galilean and Cartesian, was very generally accepted. It was therefore very natural that he should be brought into the Royal Society.

Charles Webster (1975) has defined the endeavours of Samuel Hartlib, which were (outside Cambridge) a major element in English intellectual life during the republican era, as combining Baconianism with social utilitarianism. This programme had some influence upon the nascent Royal Society, and it was a programme that, as mentioned before, held little appeal for Henry More. It moved on quite a different track from that taken by the Cambridge Platonists. Another Christ's College correspondent of Hartlib's, Thomas Smith, warned him that in Cambridge the brighter men were all so taken up with Platonism 'or other high and aery speculations of Divinity or Philosophy that they will scarce vouchsafe to cast a glance on' any new invention or indeed anything outside their own concerns (Webster 1969: 363). More, too, in his first letter to Hartlib (December 1648), used very disparaging words concerning 'men who dig up and droyle like blinde molewarpes in the earth' in search of 'usefull experiments for the discovery of Nature', without benefit of those intellectual qualities that alone make 'the close and subtill deduction of reason' possible: 'it would make a dog laugh, to think, how highly these low Spiritts, [are] commonly conceited of themselves, and are more easily proud and contemptuous than those, of far higher and more enlarged facultyes.' Surely More, kindliest of men, would not have written these words (for which he apologized later) had he been aware that Hartlib's principal effort was to bring to social fruition the work of just such 'low Spiritts' (Webster 1969: 363; 365).

One of Hartlib's devoted associates, soon to win great distinction and later to be a notable Fellow of the Royal Society, stepped in as the champion of empirical inquiry. This was William Petty, to whom More responded in March 1649. Petty was clearly stung by More's allusion to 'slibber-sauce

Experiments' and retorted with the practical uselessness of Cartesian philosophy: what had his dioptrics done for the telescope, or his physiology for medicine? Descartes might be the Prince of Philosophers but these were no more than 'Nutshells'; one fertile experimenter like Cornelius Drebbel was worth more than a crowd of them. In his witty answer More says there must be 'rome enough, for both these Geniuses to pitch their booths in, and open their shops, and show their variety of wares'. He himself counts experience as the best of knowledge, but 'the first and most generall principles of Nature have more of Divinity and Majesty in them than ever to suffer themselves to be Hermetically imprison'd in some narrow neck'd glasse, or like a Jack in a box to astartle the eyes of the vulgar at the opening of a Lidd.' As for any way of 'advancing of learning sett on foot, that experimentall knowledge forsooth may flourish like Ivy and *Leucoium* growne out of the cracks and breaches in the walls of ruin'd monasteries', he says, 'may mine eyes never see that experiment.' For several years Hartlib sought to enlist More under his banner, without ever succeeding; the epistemological differences between the idealist and the empiricist can never be reconciled and it was impossible for More to go down any path but his own. Unable to look ahead twenty-odd years to his intimate friendship with Francis Mercury van Helmont, from whom he would beg remedies for ailing kinsfolk, More particularly complained of the Paracelsans who loved 'to be tumbling of and trying tricks with the Matter (which they call making Experiments) ... This is that that commonly makes the Chymist so pitiful a Philosopher' (1969: 367–8; 369–71; Webster 1975: 147).

It is no wonder, therefore, that later attempts to convince More by the testimony of experiments that spirits are not the agency of quite ordinary natural phenomena were doomed to failure. How could the dubious and fallible evidence of the senses prevail over the clear and certain light shining from reason and Scripture? More was quite willing to propose or to repeat experiments himself, but only in order to confirm his assurance that those who quoted experiments against him were mistaken.

But it is worth noting that when More rejects the hypothesis of the 'spring and weight of the air' adopted by the pneumatic experimenters Torricelli, Pascal, Robert Boyle and others, it is not merely because of his preference for an animistic interpretation. He appeals to 'experiment' on his side. In the extended version of *An Antidote against Atheism* (1662) he argued that the atmosphere has neither weight nor spring because a pat of soft butter is not flattened by the incumbent weight, nor is the butter-milk expressed from it by the spring (1662: 44–6)![9] Clearly, More had not read Pascal who first answered parallogisms of this kind, and did not fully grasp the content of the hypothesis that he was seeking to refute. Like Newton in a different context, More denies that appeal to a non-mechanical principle of explanation is a 'vain Tautology, or the mere saying a thing is so, because it is so, but a distinct Indication of the proper and immediate cause thereof'. To look for the ghost in the machine is not 'to take sanctuary in an *Asylum of Fools'*; and further, to 'conclude that to be by *Sympathy,* that we can demonstrate not to be by mere *mechanical Powers,* is not to shelter a man's self in the *common Refuge* of *Ignorance,* but to tell the *proximate* and *immediate cause* of a *Phenomenon,* which is to philosophize to the hight' (1662: xv). Alas, Popper's criterion of falsifiability was still far into the future, but what had happened to the notion of *docta ignorantia*?

Was More merely cavilling when he drew a distinction between the mechanical philosophy which attached motion to matter once for all, in the Cartesian manner, and hence solved the problem of activity in the phenomena of Nature by the assertion that it was all a matter of the transference from particle to particle of a quantity of motion which remained (in sum) forever constant in the universe, and that other mechanical philosophy (favoured by More himself) which presupposed the existence in the universe of new sources of activity, or rather one should say of constant sources of activity, not inherent in matter itself? By no means. The conceptual distinction made by More is real, important and fruitful. There are real difficulties – hardly appreciated by More it is true – in the way of Descartes's assumption that the quantity of motion in the universe is constant. If it is *not* constant (as Newton, for example, believed) then fresh motion has to be recruited from somewhere. Further, it is difficult to conceive that all activity (motion) in the universe is the product of immediately antecedent motion (activity), as Descartes's system requires, for if in certain circumstances (e.g. by the compression of a spring) motion is merely rendered latent, it also seems that in others (e.g. by chemical reaction) it may be created *de novo*. There are other difficulties with the continuity of motion required by Descartes's system. More was quite justified in claiming that while gross phenomena certainly result from changes at the microcosmic level, these changes cannot be proved to be self-generating and self-perpetuating. They may have other causes, be these spiritual (as More himself believed) or virtues, powers and forces somehow inherent in matter or space (as Boyle and Newton believed).

Unfortunately, More was not a good judge of where to draw the boundary between mechanical and spiritualist interpretations of natural phenomena. In his writings, it seems that the power of spirits as active agents is so great that there is no room for purely mechanical explanations at all. As we saw in a previous chapter, More thought that gravity certainly could not be explained by mechanical causation, and so by inference (and in accord with More's own statements) the phenomena of hydrostatics could not be accounted for by merely mechanical causes either. (Though More did not *need* to suggest that Archimedes and all subsequent mathematical philosophers treating of hydrostatics had been in error, since each of these philosophers simply took the action of gravity for granted in order to be free to discuss the behaviour of fluids and floating bodies, More's language seems to insist upon the need for an altogether more animistic interpretation than theirs.) The fact is that in his later years More lost all interest in the study of the *immediate* causation of events in the universe, to which the concern of natural philosophers is usually limited, and concentrated his whole attention upon their ultimate final or metaphysical cause, which he found in the Spirit of Nature. It would have been surprising, therefore, if More had not also challenged the mechanical philosophy as applied to the phenomena of aerostatics, or pneumatics.

The problems, and the physical experiments giving rise to them, were far from all new. The ancients recognized that *Natura abhorret vacuum*: a strong arm is required to pull a well-fitting piston out of a closed cylinder, and the violent inrush of air as it emerges is audible. In the syringe-pump, water rises unnaturally to fill the void space, and so on. Ctesibios is said to have exploited such effects in his father's barber shop before 300 BC. Hero of Alexandria was the last and best-known exponent of the ancient pneumatic tradition, some 400 years later (Boas 1949; Hall, M. B. 1971). Just before the middle of the seventeenth century it was realized that a quantitative *limitation* to Nature's resistance to a vacuum can be exactly demonstrated; in the simple barometer (as we now call it) the level of the mercury in the tube never much exceeds 30 inches, however long the tube may be; the height of the mercury varies a little from day to day, and falls steadily as the barometer is carried higher above sea level. Moreover, when the less facile construction of a barometer with water instead of mercury was attempted, the length of the water-column standing in the tube was found to be as many times greater than 30 inches, as the density of water is less than that of mercury – it reached some 30–35 feet. All this suggested to the acute investigators that the weight of the fluid column was counter-balanced by another weight pressing downwards: that of the atmosphere itself, nearly steady at any one place but reducing at higher altitudes where the sea of superincumbent air became more shallow.

It was one thing to declare that air, like dense fluids, possesses weight, but this notion alone was not enough to account for all the experiments, especially those made with very primitive air-pumps in the 1650s. These indicated that air is also elastic, and this view was confirmed by the opposite phenomena of the compression of air. As it happened, the air-gun (of which the infant's pop-gun is the basic form) had been devised at about the same time as the barometer. Forced by a pump into a strong metal sphere, air could be reduced to a small fraction of its normal volume; on release, its elastic force was enormous. Equally elasticity was manifest in extracting air from a vessel, which was like stretching a metal spring. Combining these ideas with notions of fluids, it was not difficult to explain, for example, why a barometer still works when immersed in water, indeed why everything on Earth is at atmospheric pressure.

Robert Boyle had settled himself at Oxford towards the end of the republican period in order to devote himself to his 'grand design of promoting experimental and useful philosophy' (and to his religious studies) in peace, and there he had Robert Hooke construct for him an effective laboratory air-pump, with which they were able to explore the properties of the partial vacuum that it produced and of the elasticity of air. Presumably Boyle had read something of the earlier writings of Stevin, Torricelli and others but the immediate background to his own investigations was in Gaspar Schott's *Mechanica hydraulico-pneumatica* of 1657 (Boas 1952: 417–22; Shapin and Schaffer 1985), for Schott had described the experiments of Otto von Guericke at Magdeburg a year or two previously; his very crude air-pump had demonstrated publicly the very great effects of atmospheric pressure and elasticity. No one before Boyle had thoroughly explored the physical (and indeed, chemical) properties of the atmospheric 'vacuum'.[10]

After a fairly short but intensive period of experiment – in its way, something wholly new in the history of science – Boyle published *New*

Experiments Physico-Mechanical touching the Spring of the Air, and its Effects, made, for the most part in a New Pneumatical Engine in 1660. Very early in the treatise, Boyle 'insinuate[s] that notion, by which it seems likely, that most, if not all of them [the experiments to be described] will prove explicable'. This is the notion of the elasticity or spring of the air:

> our air either consists of, or at least abounds with, parts of such a nature, that in case they be bent or compressed by the weight of the incumbent part of the atmosphere, or by another body, they do endeavour ... to free themselves from that pressure, by bearing against the contiguous bodies ...
> (Boyle 1772, I: 11; see also Henry 1990)

Boyle's mechanical model is simple. The air is reducible to single particles; each particle is elastic, therefore the mass of the air is elastic. Each and every particle possesses a 'power or principle of self-dilatation' when released from compression, to an indefinite extent; each is in principle like a little spring of coiled steel, or for that matter a blown-up bladder. Boyle does not say where the power or principle comes from, nor does he need to; it is sufficient for his purpose that the analogy he uses be understood by his reader. (If we say that *this* lawn-mower is propelled by an engine like that of a motor-cycle, and *that* by an electric motor like that in a Hoover, one does not need a degree in engineering to understand the contrasts and comparisons intended.) The atomic structures that account for elasticity Boyle could resign to future discovery. He only needed as his theoretical assumption to claim that air, like other substances, possesses this property: 'if I have shown, that the phaenomena, I have endeavoured to account for, are explicable by the motion, bigness, gravity, shape and other mechanical affections of the small parts of liquors, I have done what I pretended' (1772, III: 608).[11] Boyle's point is that just as a mechanical explanation is not invalidated by ignorance of the precise shape, weight, speed and so on that the particles really have, so he does not need to go beyond the postulation of elasticity in the particles (understood sufficiently by analogy) to search for the causes of this elasticity. Granted the assumption, the phenomena can be solved mechanically. It is not even necessary (Boyle recognizes) that each air-particle be an individual little spring, though he thinks this the simpler view to take; the elasticity of air might be caused kinetically, as Descartes supposed, by the random or oscillatory motions of the particles.

It was also fairly easy for Boyle to demonstrate that in many of the experiments he made the spring of the air is far more significant than its dead weight, and in this of course air differs greatly from a liquid. The chief effect of the weight of the atmospheric layer about the Earth – perhaps as much as 50 miles deep – is to compress the elastic particles here at the surface 'and keep them bent'. When the pressure is taken off inside the exhausted vessel by the pump, the air can expand to at least 150 times its normal volume. Employing the concept of the normally compressed atmosphere Boyle was able to argue strongly against the traditional beliefs that suction is a positive or attractive effect – whatever sucks drawing the fluid or other matter towards itself – and that Nature abhors a vacuum. Suction or the diminution of atmospheric pressure can work effects only negatively: the positive agent is the springiness of particles of air (1772, I: 22, 70–7).

More seized his first opportunity to record his detestation of such ideas, which he understood as attributing to matter an inherent power of activity that it did not possess. He and Boyle expressed mutual esteem, but their minds and temperaments were far apart (in theological as well as natural philosophy (Henry 1990)) and it does not seem that their personal relationship was close – for More was never in Oxford, and rarely in London, especially in later life. More's hydrostatical and pneumatic experiments at Ragley, of which no details are known, were obviously undertaken to refute Boyle's theory, though More does not seem to have brought forward any experimental results of his own in rebuttal of Boyle's; the latter (for so gentle a man) manifested a degree of impatience with More's attempt to set himself up as a rival authority seeking to undermine all Boyle's accomplishment in substantiation of the mechanical philosophy. In this aim More had several coadjutors, none congenial to him, for he liked the philosophy of Thomas Hobbes and Sir Mathew Hale as little as he approved of Boyle's espousal of the mechanical philosophy. But we need not consider here his opinion of these 'allies'

Soon after the publication of Boyle's *New Experiments*, indeed amid the controversy concerning his theory, More was engaged in reworking his books for the collection of his *Philosophical Writings* published at Cambridge in 1662. He attached great importance to this lengthy reformulation of his thought, writing to Lady Conway (15 March 1662):

I have completed all things so exquisitely to my minde that I would not for all the world but that I had had this opportunity of revising them, so fond am I of the fruits of my owne minde, which yet I think I should not be, did I not hope that they will be very serviceable to the World in their chiefest concernes. (Nicolson 1930: 199)

When reconsidering *An Antidote against Atheism* for this collection, More added to it pages specifically condemning the notions of atmospheric weight and pressure. The facts (he thought) were against them: on the contrary, the *horror vacui* is a manifestation of the Spirit of Nature, a paradoxical spiritual agency working *against* mechanism: 'There is a *Principle* in the World that does tug so stoutly and resolutely against the *Mechanick* Laws of *Matter*, and that forcibly resists or nulls one common Law of Nature, for the more seasonable exercise of another' (1662: 43–6). (Whatever the precise sense of the last clause, it is perhaps just worth remarking here the analogy with the later notion of a Vital Principle opposing the normal laws of physics and chemistry in organisms, laws by which death and decomposition follow immediately after the cessation of the Principle.) Presumably More meant that *his* Principle, the Spirit of Nature, is able 'unnaturally' to raise a weight, in order to gain the higher object of filling the vacuum. To him, 'the sucker of the Airpump's drawing up so great a weight' was a notable demonstration 'that there is a substance distinct from Matter in the World', capable of purposive action.

If Boyle was soon aware of these criticisms of More's, he either chose to take no notice of them or felt that they were sufficiently dealt with in his answers to other critics. The two men met occasionally during the 1660s, partly because Lady Conway was eager to test the efficacy of one of Boyle's chemical remedies; More would call on Boyle when he went to town after Boyle's removal there in 1668. In thanking Boyle for a copy of his *New Experiments and Observations touching Cold* in 1665, More congratulated Boyle on his 'so faithful and multifarious experiments … They certainly are of far greater consequence … than the framing of any hasty hypotheses, though witty, and within some circuit of considerations, pretty coherent' (Boyle 1772, VI: 512–13),[12] and added other compliments, signing himself 'your most humble and affectionate friend and servant'. For all that, Boyle's *Hydrostatical Paradoxes* of the following year did not please More, who was not 'altogether satisfyde that his paradoxicall Inferences from the experiments are true. There will be a Spiritt of Nature for all this' (Nicolson 1930: 269). The two men were able to agree, however, on faith in the powers of Valentine Greatrakes, and lack of faith in the merits of Henry Stubbe (1930: 273).

Not until 1669 did Boyle most delicately record 'a very learned writer's' objections to his own natural philosophy, 'as well confidently as very civilly proposed', and hint that an answer to them might be provided in due course (1772, III: 274–5).[13] Before that had been prepared, More made a fresh and stronger attack upon Boyle's mechanistic interpretation of his experiments in *Enchiridion Metaphysicum* (1671), an attack that seemingly caused Boyle some annoyance. Others seem to have expected it: at any rate, John Beale wrote to his friend the Secretary of the Royal Society, Henry Oldenburg:

> Did I not foretell yu, wt was to be expected from Dr H. M. His confidence is as strong as Enthusiasme; & yet yu see wt he does. As if he had a mind to draw a suspicion, or (at leaste) to rayse ye style of Infamy agst ye RS and candid Experiment, as to be so Magicall as to call in ye ayde of Spirits & Angells. (Hall and Hall 1971, VIII: 120; Henry 1990)

Beale thought that *Enchiridion Metaphysicum* merited a 'chastisement' from Boyle; himself a somewhat eccentric Fellow of the Royal Society, Beale would surely have been upset by More's mockery of his own notion that angels might sport beards, if it came to his ears!

Enchiridion Metaphysicum, which More had completed by the summer of 1670, is perhaps the most important of his works with respect to natural philosophy, if only because it directly attacked the ideas put forward by leading figures in the Royal Society, and provoked their counter-refutations of the Hylarchic Spirit. The book propounded few metaphysical notions that More had not already enunciated, the most important innovation being the clear suggestion that Space and God are one. More maintained that Descartes was wrong to contend that those spaces in Nature commonly regarded as vacuous are all filled with that corporeal material substance called matter:

> I on the contrary when I have so manifestly proved that the internal space or place [*Spatium sive Locum*] is really distinct from matter, I conclude that it is for that reason a certain incorporeal substance or spirit, just as the

Pythagoreans formerly thought. And so through that same gate through which the Cartesian Philosophy seemed to intend to exclude God from the world, I on the contrary (and I am confident that success will be vouchsafed me) strive to reintroduce Him. And this infinite and immobile extension appears to be not only real but divine.

More then lists twenty qualifications of divinity which, he says, are also true of absolute space: 'One, Simple, Immobile, Eternal, Complete, Independent, Existing of Itself, Existing in Itself, Incorruptible, Necessary, Immense, Uncreated, Uncircumscribed, Incomprehensible, Omnipresent, Incorporeal, Permeating and Completing All Things, Essence, Activity, Pure Act' (1671: 69). At a stroke, More seems not only to obliterate any distinction between the Hylarchic Spirit and the divine essence, but to dispense also with the distinction between Spirit and universe. Spirit is not merely the active force in Nature giving motion to matter, it is the immobile, perpetual, unchanging context within which matter is moved – an absolute space or extension which has acquired an ontological rather than a merely geometrical significance. God is everywhere, and only his unceasing action gives inert matter any function or effect.

Besides being, like everything More wrote, 'a Demonstration of Incorporeal Beings', the *Enchiridion* was 'as well a confutation of the Mechanick Philosophy', as he told Lady Conway (Nicolson 1930: 303, 6/8/1670). The latter part of the book indeed contained forceful attacks not only on Hobbist and Cartesian physics for their materialism and actual failure to achieve their objectives in explaining phenomena, but on the no less mechanical physical theories of Robert Boyle and Robert Hooke. Chapters XII and XIII of *Enchiridion Metaphysicum* made it impossible for Boyle to prolong his former silence: he found that More had endeavoured

> to make my opinions appear not only untrue, but irrational and absurd, [so] that I feared his discourse, if unanswered, might pass for unanswerable, especially among those learned men, who, not being versed in hydrostaticks, would be apt to take his authority and his confidence for cogent arguments. (Boyle 1772, III: 597)

Boyle's answer came in the miscellany of papers, a bibliographer's nightmare, that Boyle published in 1672 under the title *Tracts ... containing New*

Experiments touching the Relation betwixt Flame *and* Fire ... (*with much more*) (1772, III: 596–628; Fulton 1961: 70–1). First, Boyle argued that it was perfectly proper to assign weight to the particles of a fluid or air, without explaining the cause of gravity, weight being a fact known to us by experience. Second, Boyle pointed out, pressure acts equally in all directions in an elastic fluid like air, as it does in an inelastic fluid like water (as More might have gathered from Archimedes, but Boyle does not say so). More's third quibble is likewise founded only on ignorance of hydrostatics. Fourth, More's contention that the massive atmospheric pressure alleged by Boyle would crush everything on Earth is refuted by consideration of the omnipresence of pressure and the absence in bodies of vacuities impenetrable to the air, even in soft bodies; so, if something as weak as the film bounding a bubble is neither wrinkled nor broken by the atmospheric pressure, it is because the air within the bubble is already compressed to the same extent. (In a later experiment Boyle found that a fly could withstand without detriment a pressure of more than 20 inches of mercury (1772, III: 606; 623).) Fifthly, Boyle showed that More's 'refutation' of Boyle's argument that air-pressure contributes to the mutual adhesion of two plane surfaces was equally false; More's thought-experiments on his own behalf were readily disposed of.

Boyle next turned to More's criticism of *Hydrostatical Paradoxes*. He opened with a rather long statement of his general position, concluding that More's introduction of spirits or angels was not so much disproved by his own kind of natural philosophy as rendered redundant. He declared his own belief in a divine Creator – a belief he had already claimed for Descartes as well as himself – and went on:

> all that I have endeavoured to do in the explication of what happens among inanimate bodies, is to shew, that supposing the world to have been at first made, and to be continually preserved by God's divine power and wisdom; and supposing his general concourse to the maintenance of the laws he has established in it, the phaenomena, I strive to explicate, may be solved mechanically, that is, by the mechanical affections of matter, without recourse to nature's abhorrence of a vacuum, to substantial forms, or to other incorporeal creatures. And therefore, if I have shewn, that the phaenomena, I have endeavoured to account for, are explicable by the motion, bigness, gravity, shape, and other mechanical affections of the small parts of liquors, I have done what I pretended; which was not to prove, that no angel or

other immaterial creature could interpose in these cases; for concerning such agents, all that I need say, is, that in the cases proposed we have no need to recur to them. And this ... has been by almost all the modern philosophers of different sects thought a sufficient reason to reject the agency of intelligences, after Aristotle, and so many learned men, both mathematicians and others, had for many ages believed them the movers of the celestial orbs. (1772, III: 608–9)

Another thought-experiment alleged by More – that discussed above of a disk of wood floating upwards in a cylinder filled with water only slightly exceeding the disk in diameter – was then disposed of; correctly interpreted along Archimedean lines it did not support More's opinion that a volume of fluid within a fluid exercises no downwards pressure, and so the pressure within a fluid does not increase as the depth below the surface is greater. Further (Boyle insisted) actual trials had proved that in the ocean the pressure at great depths became enormous. Boyle repeated that there is no need to 'fly' to the idea of an incorporeal director or Hylarchic Principle

> since such mechanical affections of matter, as the spring and weight of the air, the gravity and fluidity of the water, and other liquors, may suffice to produce and account for the phaenomena, without recourse to an incorporeal creature, which it is like the Peripateticks and divers other philosophers, may think less qualified for the province assigned it, than their fuga vacui, whereto they ascribe an unlimited power to execute its functions. (1772, III: 627)

After demonstrating that the substance of More's critical objections vanishes away in the light of long-accepted hydrostatic principles, Boyle remarks, as well he might, that 'it is not necessary that a great scholar should be a good hydrostatician,' and he does not forget to reiterate the metaphysical argument that belief in a deity does not logically entail belief in all sorts of subordinate spirits. Most of the 'heathen philosophers were convinced of the being of a divine architect of the world ... wherein they believed things to be managed in a mere physical way, according to the general laws, settled amongst things corporeal' without the need to invoke a Spirit of Nature.

Looking ahead to the next generation of philosophers, it is clear that Boyle, when opposing More, was much closer to the later position of Leibniz than he was to that of Newton. If God, providence or some other spiritual agency should intervene in the course of natural phenomena, this intervention must be counted as miraculous rather than as a routine event. The usual operations of Nature must follow a lawlike course, and that course must be translatable in our minds into mechanical cause-and-effect relationships; not that we can as yet *define* the mechanical properties involved (those of invisible, inaccessible particles, etc.) still less analyse completely the 'powers' of matter (such as elasticity), but we must believe that mechanical relationships do follow consequentially from the unalterable structure of things that we call material. In other words, Boyle (like Leibniz) is a unitarian, while More (like Newton sometimes) is a dualist; for the latter *one* creation, that of matter, is not enough. There had to be a separate creation of spirit to make matter active.

More was not the first metaphysician, nor the last, to be, as it were, ejected from a scientific debate because of his professional incompetence. In More's own time such intellectually distinguished philosophers as Julius Caesar Scaliger and Thomas Hobbes were summarily convicted of ignorance in geometry by professional mathematicians. It was more unusual in More's case that the charge of incompetence related to experimental science – classical as well as recent. 'In these contests with his adversaries Boyle constantly pointed to the boundary between the community of exper- imenters and external critics. Insofar as More or Hobbes failed to produce fresh matters of fact or reproduce those of Boyle, they broke the rules of experimental dispute.' A metaphysician who denied the reality of the air's spring or water's weight in water could not be regarded as a competent member of the scientific fraternity, and therefore (since they rested upon these untenable denials) his metaphysics could not be taken seriously. 'The air's spring was by now the mark of membership of the experimental community.' And Boyle also makes it clear that a proper understanding of the basic theoretical ideas of a science, and skill in applying them to the solving of phenomena, are essential prerequisites to participation in debate (Shapin and Schaffer 1985: 222; 223).

Henry More knew of Robert Boyle's intention to publish an answer to *Enchiridion Metaphysicum* by May 1672, when he told Lady Conway that Boyle did not 'take my dissenting so from him in publick so candidly as I had hoped, which I am very sorry for'. He 'thought Philosophy had been free', and would have thanked anyone who pointed out an error of his own – a nice piece of conceit, since Boyle was far from admitting that he *had* fallen into error (Nicolson 1930: 358, 11/5/1672)![14] At the end of this year More wrote to Boyle – presumably after reading Boyle's as yet unpublished *Hydrostatical Discourse occasioned by the Objections of the Learned Dr Henry More* – in justification of his personal crusade against atheism (Boyle 1772, VI: 513–15).[15] His letter is a relatively succinct statement of a position More asserted again and again. Descartes's philosophy had been taken over by the 'atheistical gang' who openly proclaimed that God's defendants, with More at their head, were a 'company of fools'. Chief of the atheists (by implication) was Spinoza in Holland, 'a *Jew* first, after a *Cartesian*, and now an atheist … supposed author of *Theologico-Politicus*'. More had to prove 'how prejudicial *Des Cartes's* mechanical pretensions are to the belief in God' and that his philosophy 'is the pillar of many of those men's infidelity, and of their atheism'. He had himself leaned to Cartesianism, though always with the proviso 'that this mechanical way would not hold in all phaenomena', and for that reason 'what could I have done less, than declare my sense of the *Cartesian* philosophy, if it be credited.' More then emphasized repeatedly his own consciousness of his great success in 'demonstrating the necessity of incorporeal beings' even in this age when 'the notion of spirit is so hooted at by many for nonsense'. The failure of the mechanical philosophy to solve even the easiest and simplest of phenomena he has proved 'irrefutably' and he is 'unspeakably confident' of his result, resting upon 'invincible arguments'. This success brings a satisfaction to his own mind 'which I value more than any thing this world can afford me'.

Still apparently heedless of the fact that with every slash at Descartes he gave a back-cut to Boyle himself, More went on (rather unreasonably) to complain that he could not have foreseen that Boyle would so dislike his book, 'doing no other to him, than what, if it were done to myself, I should take no offence at'. As for the experiments described by Boyle (which he employed for his own purposes) he could not but mention and make use of them in his own arguments because they came so aptly 'as a main pledge of the tenets I so much contend for':

> that the phaenomena of the world cannot be solved merely mechanically, but that there is the necessity of the assistance of a substance distinct from matter, that is, of a spirit, or being incorporeal, as I must confess I do conceive

— 193 —

most firmly to have concluded from these experiments; for which I ever hold the world and myself obliged to you.

Indeed, More went on, if he were to find himself mistaken in his anti-Boyle inferences from Boyle's experiments, he would never trust reasoning in philosophy again, 'but altogether endeavour after as perfect innocency as I can'.

What Boyle's opinion was of this last euphoric attempt to convince him that More's interpretation of the experiments was far superior to Boyle's own, we shall never know, since the *Discourse* was surely written long before Boyle read More's letter. It could only have reinforced his opinion that More's confidence in the reality of his spiritual agencies far outran his competence in scientific argument. All that was needed by way of rebuttal had already been said in the *Discourse*. In due course Boyle sent More a copy of the *Tracts ... touching ... Fire and Flame* and when they next met (in August 1673) at the booksellers' in Paul's Churchyard they saluted each other very kindly without adverting to the content of the volume. The passage of time did nothing to soften Boyle's feeling that More, good man as he was, had a bee in his bonnet about spirits; when More in 1676 again published on the subject of hydrostatics in *Remarks upon two late ingenious Discourses*, directed against George Sinclair's criticisms of himself, he earned from Boyle (according to Lady Conway) only the sour comment that More had better not have printed it, since he was mistaken in all his experiments. To which More, sticking to his artillery as firmly as Boyle did, retorted that Boyle was 'pinched' by his agreeing with Sir Mathew Hale in the 'exploding that monstrous spring of the ayre'. After this, there was no more dispute between the two men, but we have no evidence of any further friendly contact either (Nicolson 1930: 371; 420; 423).

There can be no two opinions about the verdict of the Fellows of the Royal Society upon Boyle's detractors, whether Linus, Hobbes, Hale or Henry More. Robert Hooke, maker of the air-pump, John Flamsteed the astronomer, Henry Oldenburg (Boyle's assistant in his literary labours, who expected to translate *Fire and Flame* into Latin) and John Wallis the mathematician are all known to have supported Boyle's 'spring of the air'. Wallis wrote that 'when all hath been sayd that can be on both sides, it must in this (as in that of ye Copernican hypothesis, & the Circulation of the Bloud), be left at last to ye Readers pleasure, with whether side to join'

(Hall and Hall 1977, XI: 37; 168). As Max Planck remarked centuries later, only death can convert the last inveterate opponents to new ideas. Oldenburg sent some copies of More's books to his correspondents overseas: one, Jean Baptiste Duhamel, future historian of the Académie Royale des Sciences in Paris, refers to More as one of those persons 'qui parlent de toutes choses avec un air decisif, et qui veulent que toutes leurs pensées passent pour des oracles, et pour des demonstrations'. He thought Boyle's reply to More strong and judicious – who would want to destroy so well founded a doctrine as that of the air's elasticity? Duhamel doubted whether other Parisian philosophers would bother with More's books (Hall and Hall 1975, X: 21–2). In the *Philosophical Transactions*, Oldenburg printed a neutral account of *Enchiridion Metaphysicum* which, nevertheless, invited the reader's incredulity with respect to More's Spirit of Nature. His account of the book is fair but cold. Oldenburg clearly did not welcome More's criticism of Descartes's treatment of motion, nor his idea 'that that Space or Internal Place is really distinct from Matter, and is an Incorporeal Spirit', nor More's concept of 'this immense and immaterial space and substance as some Representation of the Divine Essence'. Further, Oldenburg opines that the reader will be surprised to learn from this book that mere mechanical causes are inadequate to account for gravity, or the phenomena of pneumatics and hydrostatics, of optics and meteorology, and many other things. When the *Remarks* came along five years later Oldenburg dismissed them with references to Boyle's *Hydrostatical Discourses* and Wallis's *Discourse on Gravity and Gravitation* (1675), for these works would teach the reader how to judge More's new book (Oldenburg 1671: 2182–84; 1676: 550).

Robert Hooke borrowed and read More's *Remarks* on 25 January 1676 and a fortnight later began a rejoinder to them which he delivered to the Royal Society as a Cutlerian lecture two days later: there was a 'great auditory' (Robinson and Adams 1935: 214; 216). The *Remarks* presumably caused Hooke to look into *Enchiridion Metaphysicum* as well, for this became the chief target of his criticism in the published version of his lecture, in *Lampas* (1677) (Gunther 1931, VIII: 182–95). Hooke wonders that More should deny the gravitation of fluids (including air), so long 'demonstrated by Stevinus ... and ... by many more modern Authors who have writ most learnedly and clearly thereof'. Instead, More invokes 'an *Hylarchick Spirit*, which at command acts and performs whatever is necessary to solve all the

Phenomena of Mechanical, Hydrostatical, and, in a word, all Physical motions and effects'. Hooke admits that, like More, he does not subscribe to Descartes's mechanical explanation of gravity, but denies that this is (as More declares) absurd. And why should not a fluid mass gravitate exactly as the identical mass, solidified, would do? And that there is increasing pressure at greater depths in a fluid is proved (Hooke ingeniously notes) by the fact, experimentally established, that the velocity of efflux of water from a hole in the side of, say, a barrel is proportional to the square root of the depth of the hole below the water level. (This, Hooke mistakenly adds, follows from the increase of acceleration of a movement in the ratio of the square root of the force applied, in this case pressure, which varies directly with the depth.)[16] Hydrostatics therefore agrees with the laws of mechanics, and the Doctor offers nothing convincing against this agreement, either in the *Enchiridion* or the *Remarks*, 'but what seems grounded upon some pre-conceived Notions and Hypotheses which I cannot understand; and I cannot see how he can avoid acknowledging this, to be a Mechanical motion, if at least he will allow any Mechanical motion at all ...'. If we suppose fluids to gravitate like other bodies we are enabled to solve all the problems of hydrostatics, and it is needless

> to have recourse to an Hylarchic Spirit to perform all those things which are plainly and clearly performed by the common and known Rules of Mechanicks ... which do not perplex our minds with unintelligible Ideas of things, which do no ways tend to knowledge and practice, but end in amazement and confusion.

Hooke further observes, acutely, that a natural philosopher seeking to explain phenomena has to choose between law and the absence of law, or caprice. If the Hylarchic Spirit is capricious and unconstrained, is it to be moved by appropriate 'Charms and Incantations' inciting the Spirit to be more or less active, to make the flow greater or less in a given situation? On the other hand, if the Spirit observes inviolable laws, 'what were we the better or the wiser unless we also know how to rule and govern this Spirit?' Further, if the rule of law and the constant regularity of cause and effect be abandoned in natural philosophy, the investigation of phenomena is discouraged:

For if all things be done by an Hylarchic Spirit, that is, I know not what, and to be found I know not when or where, and acts all things I know not how, what should I trouble my self to inquire into that which is never to be understood, and is beyond the reach of my Faculties to comprehend?

Hooke had read far enough in *Enchiridion Metaphysicum* to discover in chapter XIX 'some Animadversions upon an Explication of Colours which I did formerly publish in my *Micrographia*, from the confutation of which [More] endeavours to assert this Hylarchic Spirit'. Now More had placed himself in the curious position of arguing at the same time both in favour of Descartes's mechanical, globular hypothesis of light and his own Hylarchic Spirit. Hooke finds no difficulty in pointing out the technical shortcomings of More's account, and the impossibility of Descartes's hypothesis explaining the ring coloration of thin plates Hooke had himself first observed. If More were correct in his defence of Descartes, then 'the Image from a Looking-Glass must be returned coloured, and the same also from a plain sided Prisme, ... But this is contrary to Experiment ...' He now charges More with employing his own arguments against Cartesian physical optics in *Micrographia*, without acknowledgment, in order to show the inadequacy of a corporeal theory of light, although he has just defended that theory against Hooke! He concludes

> I could heartily therefore have wished that the learned Doctor had made use of some other Mediums to prove the Existence of an Hylarchic Spirit, and not have medled with Arguments drawn either from Mechanicks or Opticks; for I doubt, that such as understand those subjects well, will plainly see that there is no need of any such Hylarchic Spirit; and if there be no need of it, but that all the Phenomena may be done without it, then it is probable that there is none there, for *Natura nihil agit frustra*.

And rather wickedly Hooke advises More to employ

> Arguments drawn from subjects we less perfectly understand, as from the generation, nutrition, vegetation, and propagating of Vegetables and animal substances; for there the manner of the progress of Nature being infinitely more curious and abstruse ... one may more boldly assert strange things of this Hylarchick Spirit without fear of controul or contradiction, and from

whence possibly it may never lie within the power of Reasoning to banish him. (Gunther 1931, VIII: 194; 195)

Like Boyle, Hooke has pitted his professional expertise against More, and he finds the Doctor as ignorant of optics as he is of hydrostatics. It is not only the case that the appropriate experimental evidence is unfamiliar to him, or that he is unable to interpret it convincingly; the accusation is also that More has failed to study the relevant authorities (for example, Stevin and Mersenne) and so his grasp of the theoretical principles of the sciences is weak. He is unable to generalize from the laws of motion to the behaviour of fluids, and fails equally with refraction. To use the language of Shapin and Schaffer, Hooke pushes More even further out of the scientific community than Boyle had done: Hooke treats him as wholly incompetent to enter into scientific discussion and as constituting, indeed, a danger to the modern scientific movement. Descartes might have made mistakes, but he was on the right side. Boyle's rebuke to More is tempered by a consciousness that More was (in a misguided way) a participant in Boyle's own struggle to serve natural religion, to create an impeccably Christian natural philosophy. But Hooke, though a natural-religionist too, made no such concessions.[17]

Almost as though following Hooke's sound advice, More made a third onslaught on yet another Fellow of the Royal Society, Francis Glisson (c.1597–1677), Regius Professor of Physic in More's own University, choosing as his vehicle a second *Epistola ad V.C.*, printed in More's *Opera Omnia* of 1679. The first *Epistola ad V.C.*, separately published in 1664, had marked a significant stage in More's public divorce from Descartes.[18] Glisson, a biologist and physician in the Harveian tradition, had been active in anatomical and clinical investigation in both the College of Physicians and the Royal Society. The best-known new concept associated with his name is that of *irritability* – the idea that the fibres of living substance can be stimulated, and react, independently of the nervous system and indeed of the whole organism of which the particular fibres are a part.[19] In 1672, after publishing no major work for twenty years, Glisson printed a *Tractatus de natura substantiae energetica, seu de vita naturae* ('A treatise on the energetic nature of substance, or the life of nature'), in which he argues that there is no such thing as 'dead and senseless' matter (Marion 1880; Henry 1987: 17–22).[20] True, only living matter is capable of self-motion; nevertheless inorganic matter too possesses a kind of life which is realized in its properties. 'The difference, for Glisson, between what was usually considered to be non-living and living matter was simply one of organization' (Henry 1987: 21). By a kind of multiplication or intensification, the dimmest level of being, the inorganic, could acquire a higher level of life so becoming organic, attaining the vegetable life of plants and then the sensitive life of animals. Glisson thought that the higher forms of organic being could not exist unless there were first life in the inorganic matter from which they derive ultimately:

> Matter contains all forms, and consequently every kind of material life within itself potentially (*in sua potestate*). Material souls are themselves matter of a sort, and modify its primitive life in a variety of ways ... For life has different characteristics in plants and animals ... and is diversified by a marvellous variety of organization. However, unless the original life of matter provided the foundation for all these modifications, they by themselves would not impart life. (Glisson *Tractatus* (1672), *Ad Lectorum*, §9, quoted Marion 1880: 34)

Glisson agreed with More that spirits have a substantial existence which could be annihilated by God, but thought that their power came from an 'energetic nature', which is immortal.[21] This 'energetic nature' exists also in matter, and it is by modification of it that matter rises to the higher, organic level in plants and animals.

Glisson, like Harvey, was deeply enmeshed in Aristotelian thought and language, so that his profound analysis of the dependence of the living world upon earth, water and air and of the gradations of life (upon which Aristotle so often insists) does not come clearly through to us. His curious theory is a good deal more elaborate than these few hints suggest, from which, however, it will appear that Glisson sought a *continuity* between the non-living and the living where Descartes and More, pursuing different philosophical purposes, nevertheless agreed in positing a complete *discontinuity*. (In fact, of course, dualism is even more prominent in More's philosophy than in Descartes's, since the latter only introduced it in relation to the human soul while for More dualism is universal.) Now it was just as heinously erroneous, in More's eyes, to endow matter with a principle of activity, or spirit, of its own as to deny the existence of spirit altogether while maintaining that everything is brought about by the restless activity of matter in motion. The situation in philosophy resembles the Gilbertian point about distinctions in society: if the principle of activity, or life, or spirit (the precise term is unimportant) be supposed to be universally distributed as a concomitant of matter in Glisson's manner, then the distinction between matter and spirit loses much of its force: spirit loses its separate, loftier existence. More might obviously have joined himself with Glisson simply by redefining Glisson's *natura energetica* as his own Spirit of Nature, but Glisson's ineluctable union of *natura energetica* with matter made this course impossible for him.

Ralph Cudworth had already (it seems) taken Glisson to task – though without naming him – for being a 'hylozoist' in *The True Intellectual System* (1678) before More spoke out (so Henry 1987: 28, his quotation being from the first edition, p. 105): 'Hylozoism [wrote Cudworth] makes all Body as such, and therefore every smallest atom of it, to have Life Essentially belonging to it ... as if Life, and Matter or extended Bulk, were but two Incomplete and Inadequate Conceptions, of one and the same Substance, called Body.' With such a conception of 'plastick life' as the antecedent of the life of plants and animals, there was, Cudworth held, no need for a rational human soul nor for divine management of the universe. More was bound to share these views. He seems to have encountered the Spinozan form of hylozoism or panpsychism in the Spring of 1677, and he made this the principal target of his attack in *Ad V.C. Epistola altera* (Nicolson 1930: 429, 3/4/1677; Henry 1987: 28–38). In the pages devoted to Glisson, More did not hesitate to name the Regius Professor, but he did not repeat the mistake of involving his own interpretation of experiments in his criticism. From Henry's account of the matter it appears that More refuted Glisson by a flat contradiction and the assertion of his own Spirit of Nature. Where Glisson had found examples of the inherent activity of matter in the revolutions of the heavens, the cohesion of solid bodies, and the elasticity of certain substances, More claimed all of these phenomena as instances of the operation of the Spirit of Nature, 'which holds together and activates the whole material world'. As a more general offence, Glisson's philosophy was atheistical: 'Only those who deny God and all incorporeal substance strive to seek the origin of motion and all life that gleams in the universe in matter itself.' Henry describes More's riposte to Glisson as 'question-begging of astonishing impudence' (1987: 28–32). It is impossible not to agree.

It may not be entirely superfluous to point out that while the debates between More and Boyle and More and Hooke can properly be regarded as concerned with natural philosophy, the issues between More and Glisson were wholly metaphysical. It is impossible to conceive of any phenomenalistic test to decide between their two positions. And here, beyond the limits of argument and experiment, the Cambridge Platonist's confidence in reason breaks down and he is left with assertion as his only refuge. As Rosemary Colie has said, the Cambridge Platonists were no less aware than the extreme sectarians that 'one contributing cause to atheism might well lie in too close study of natural philosophy,' even though they also understood that natural philosophy might yield manifestations of God's eternal providence (Colie 1957: 5; see also Hutton 1984). If there were those, like the Spinozans, who took the fatal step from mechanism to atheism and who were wilfully blind to the manifestations of providence in Nature, it was proper to denounce their error and their wilfulness. The association of More's attack upon Glisson with the first of his onslaughts upon Spinoza shows well enough that the matters in contention lay far deeper than natural philosophy. More's denunciation of Spinoza and all his works, ably discussed by Dr Colie, need not further concern us here, unless to note that the Dutch philosopher's scorn for natural religion was firmly rebutted by More. Spinoza had written 'anyone who seeks for the cause of miracles, and strives to understand natural phenomena as an intelligent being and not gaze at them as a fool, is set down and denounced as an impious heretic' (1957: 90). That was to be his own fate. More in reply recapitulated all that he had written in favour of God's works and against blind mechanical necessity. He had nothing new to say, though the bitterness of his rhetoric increased. More was, in these last years, firmly set in his own pattern, which was not to change.

10

More and Newton: Space and Time

A S EVERYONE KNOWS, Isaac Newton in his *Mathematical Principles of Natural Philosophy* drew important distinctions between *absolute* time and relative time, *absolute* space and relative space, and it is hardly less generally believed that Newton derived these distinctions from Henry More.[1] And yet there is room for some doubt on the matter. In his published writings Newton barely mentioned More by name; this is a less decisive point than it might seem, since Newton was silent in public about other authors whose works are known to have been read by him, not without effect (as it is supposed) upon his own thoughts, and was parsimonious in such references as he did make to the giants on whose shoulders he claimed to stand, including Galileo and Kepler. I do not, in fact, believe that More's highly imaginative, sometimes flamboyant, style of writing would have been congenial to Newton, although a number of his books were to be found in Newton's library. One, *Tectractys anti-astrologica* (1681),[2] purports to have been More's gift to Newton. Another was one of More's pamphlets against the alchemist Thomas Vaughan. Four works are More's studies of the prophecies of Daniel and John, a subject of intense scholarly interest to Newton; one of these too was presented by the author.[3] Newton also owned a copy of the *Philosophical Poems* of 1647, which he had bound with *An Antidote against Atheism* (1653), and *The Immortality of the Soul* (1659). The copies of the *Antidote* and of *A plain and continued exposition of the several prophecies or divine visions of the Prophet Daniel* (1681) bear annotations in Newton's hand (see Harrison 1978: 195–6). The total absence of More's anti-Cartesian writings, especially *Divine Dialogues* and *Enchiridion Metaphysicum*, is noteworthy.

Even more to the point, Isaac Barrow, the first Lucasian Professor of Mathematics at Cambridge and later Master of Newton's College, Trinity, expounded in his lectures ideas similar to those of More and Newton – in fact, his expressions are closer to Newton's later formulation than are those of Henry More. If it is untrue that Barrow ever exercised a formal responsibility for Newton's education in their College, it seems certain that the relations of the youthful Newton with the Lucasian Professor were closer than those with More, though the latter was from the same county and the same school. For Barrow was a mathematician, and though when a young man Newton certainly had a considerable interest in the sort of metaphysics in which More excelled, it was not to be his major preoccupation. It might be argued – and is indeed implicitly argued by Burtt – that besides influencing Newton directly More also influenced him indirectly through Barrow. But the indebtedness of Barrow to More in this realm of thought is even less clear than that of Newton: little can be presumed on the mere fact of their presence in the same University during the same years. Barrow and More were not at all men of the same religious, political or intellectual temper, and Barrow, though a theologian like More, was given to neither mystical nor metaphysical notions. The two men no doubt respected each other while each pursued his own path. As we shall see shortly, the character of Barrow's discussion of the problems of space and time was quite alien to that of More.

Apart from the abortive affair of the Cambridge philosophical club (p. 170) the only direct meeting between More and Newton of which we know was described by More in a letter of 1680 to John Sharp, his former pupil at Christ's and later Archbishop of York. It seems that Sharp had asked More about his acquaintance with Newton, and how well they agreed in their interpretations of the Apocalypse of St John:

I remember I told you both how well we agreed. For after his reading of the Exposition of the Apocalypse which I gave him, he came to my chamber, where he seem'd to me not onely to approve my Exposition as coherent and perspicuous throughout from the beginning to the end, but (by the manner of his countenance which is ordinarily melancholy and thoughtfull, but then mighty lightsome and chearfull, and by the free profession of what satisfaction he took therein) to be in a maner transported. So that I took it for granted, that what particular conceits he had of his own had vanished. But since I perceive he recoyles into a former conceit he had entertained, that the seven Vials commence with the seven Trumpets, which I always look'd upon as a very extravagant conceit, and he will not have the Epistles to the Seven Churches to be a prophecy of the state of the Church from the beginning to the end of the world. He will also have the three dayes and half of the witnesses lyeing slain, to be three yeares and a half after their mournfull witnessing. Mr Newton has a singular Genius to Mathematicks, and I take him to be a good serious man. But he pronounces of the Seven Churches not yett having read my Exposition of them, ... concerning which another Mathematicall head, noted for that faculty in Cambridge, writ to me, that with Mathematicall evidence I had demonstrated that truth. And I do not doubt but when he shall have read my threefold appendage to my Prophecyes of Daniell, ... he will be of the same minde with myself. When yett if he be not, it will signify nothing to me... he has free leave from me to enjoy his own opinions. We have free converse and friendship, which these differences will not disturb. (Nicolson 1930: 478–9, 18/8/1680)

It is tempting to speculate that the second 'mathematical head' mentioned might have been that of Isaac Barrow, but he had been dead for three years. It is evident that the topic of discussion was far from the metaphysics of Nature. More is surely not claiming any discipleship in Newton, nor is there any hint of a personal intimacy between them.

Newton's only reference to Henry More in the writings printed during his lifetime is in an anonymous paper, 'An Account of the Book entituled *Commercium Epistolicum*', of 1715. Here Newton, writing of himself in the third person, defended his claims to the invention of the calculus against Leibniz, and at the end of the paper expounded the merits of his own cautious and experimental method of philosophizing as superior to the assertive and hypothetical method of Leibniz. It is remarkable, the 'Account' declares, that 'Mr. *Newton* should be reflected upon for not explaining the

Causes of Gravity and other Attractions by Hypotheses; as if it were a Crime to content himself with Certainties and let Uncertainties alone'; yet this has not prevented the editors of the Leipzig learned periodical with which Leibniz was associated from telling the world that 'Mr. *Newton* denies that the Cause of Gravity is Mechanical, and that if the Spirit or Agent by which Electrical Attraction is performed be not the *Ether* or *subtile Matter* of *Cartes*, it is less valuable than an Hypothesis, and perhaps may be the Hylarchic Principle of Dr. *Henry Moor*' (Hall, A. Rupert 1980: 313, reprinting *Philosophical Transactions*, 29 (1715), 223). Newton's distance from More is emphasized by the mis-spelling of his name; the passage hardly reads as though Newton was eager for others to trace any connection between his ideas and those of his Grantham predecessor. Their success in doing so would have caused them to suppose that they had laid bare the metaphysical foundations of his natural philosophy, which he was ever anxious to remove from debate: it was far safer and less troublesome to rest his theories of optics and gravitation upon 'inductions' from observation and experiment which, with mathematical principles, he took to be the firmest, unassailable bases of natural philosophy.

Nevertheless, as has long been known, More's *Immortality of the Soul* figured among Newton's early reading (Hall, A. Rupert 1948: 243; Westfall 1962). According to R. S. Westfall, Newton's biographer, More was an 'influence' throughout the series of youthful annotations and reflections entitled *Questiones quaedam Philosophiae* ('Some problems of philosophy'). The *Questiones* record not only Newton's first serious reading outside textbooks, but the start of his emancipation from the conventional ideas and lines of argument that Galileo and Descartes (besides the mathematicians whom Newton also studied) had rendered intolerably old-fashioned. Westfall suggests that More may have been the means of Newton's introduction to the 'new learning' (1971: 326; 327). At this stage, perhaps in his last undergraduate year, Newton's prime authority in natural philosophy was Descartes, but he learned from his own reflections as well as from his reading of such critical authors as Pierre Gassendi and Henry More that Descartes was very far from being impeccable. Like them, Newton from the first adopted atomism in preference to the infinitely divisible matter of Descartes, finding More's treatment of this question the most satisfactory (Westfall 1962: 174). 'The first matter must be atoms' he wrote in the *Questiones*: 'And yt Matter may be so small as to be indiscerpible the excellent Dr Moore in his booke of ye soules Immortality hath proved beyond all controversie' (McGuire and Tamny 1983: 340). But curiously enough the bulk of Newton's notes on the *Immortality* (he used no other of More's books at this time) relate to the nervous system and sensation, material that More himself must have copied from other sources. So he notes that injury to the stomach sooner causes death than damage to the brain, which 'though it take away sence yet ye hearts motion is not impeded thereby'. Again, 'A man cannot see through the hole wch a trepan makes in his head' (1983: 144; 382–3). Newton transcribes out of More a Latin quotation from Lucretius about loss of memory and the rather bizarre affirmation that 'When a snaile creepes a gale of spirits circuit from her head downe her back to her taile & up her belly to her head againe' (1983: 392; 418).

Prudently, the editors of the *Questiones* counsel against exaggeration of More's influence upon the young Newton. More 'was not a major influence on Newton's views concerning infinity and extension' and, being directly acquainted 'with many of the ancient sources of atomism on which More himself drew', Newton demonstrates a good deal of independence from More. In the theory of sensation Newton, here coming a little close to Hobbes, attaches more importance than More does to motion in the nervous system. The greatest single influence in the *Questiones*, they conclude, was that of Descartes (1983: 217; 319).

Some years after writing these notes and brief philosophical sketches Newton began another, more formal essay of whose development he completely lost control, so that he had to abandon it. Although the topic of this essay, *De gravitatione et aequipondio fluidorum* ('On the gravity and equilibrium of fluids') is purportedly hydrostatics, Newton launched at once into a lengthy attack upon Descartes's treatment of space and time, motion and rest. More is not mentioned by name, but his influence may be discerned. In broad principles, Newton's dissensions from Descartes are much like More's, and could well have been stimulated by reading the English philosopher's letters to Descartes, but we have no evidence that Newton ever did read them. Newton extends his refutation into a depth and remorseless detail of analysis – over seventeen printed pages in English – which goes far beyond anything More wrote upon these topics, and which is moreover free from theological assumptions. Newton argues less that Descartes is wrong than that his positions are illogical and absurd:

... not only do its absurd consequences convince us how confused and incongruous to reason this doctrine is, but Descartes by contradicting himself seems to acknowledge the fact. (Hall and Hall: 1962: 124)

... he seems hardly consistent when he supposes that a single motion corresponds to each body according to the truth of things, and yet that there really are innumerable motions in each body. (1962: 125)

If Descartes now says that extension is not infinite but rather indefinite, he should be corrected by the grammarians. (1962: 135)

Newton had always shared More's view – in opposition to Descartes – that something exists besides God and matter. For More it was spirit, which fills space and animates matter; for Newton it was simply empty space in which only omnipresent God exists (McGuire and Tamny 1983: 122–4). In *De gravitatione*, Newton went much further than he had in the *Questiones* in expounding a coherent theory of absolute space and absolute time, in which perhaps another influence than that of Henry More is to be discerned. Yet as Newton's own thinking matured, what he had learned from More was never forgotten or supplanted.

We have already given some attention to the ways in which More's ideas of space and motion differed from those of Descartes, and the differences have been analysed in detail by Alexandre Koyré (1957: ch. 4). In brief, Descartes's notion that motion can only be conceived relatively seemed to More pusillanimous and to lack dignity. Descartes's treatment put the measure of space and the recognition of motion in the eye of the observer – each observer must see things differently from others, and there can be no final choice between alternative accounts. More believed, on the other hand, that there must be a unique *true* account of things, and a uniquely privileged observer, who is of course God. Descartes was in error in supposing that we must attach the idea of *space* to the idea of *matter*, as though to measure were always to take the length of something and not to take the interval *between* two things:

For it is plain that not so much as our Imagination is engaged to an appropriation of this *Idea of Space* to corporeal *Matter*, ... and therefore it may as well belong to a *Spirit* as a *Body*. Whence as I said before, the *Idea* of God being such as it is, it will both justly and necessarily cast this ruder notion of *Space* upon that infinite and eternal spirit which is God.

According to More, the idea of distance has no necessary connection with the idea of matter, and space and distance would remain if all matter abstracted from the world. Thus arguing, as we have seen already (p. 188 – 9), More arrives at the identity of space and God: 'if after the removal of *corporeal Matter* out of the World, there will still be *Space* and *distance* in which this very matter, while it was there, was also conceived to lye' which must be incorporeal, penetrable and eternal, and which 'the clearer *Idea* of a *Being absolutely perfect* will more punctually inform us to be the *Self-subsisting* God'. Space is unmoved (indeed, how could it make sense to speak of space moving?) and it is also eternal and infinite, like the universe itself, which is in space but not physically coterminous with space (Koyré 1957: 135–7, quoting *An Antidote against Atheisme*: 163ff).[4]

By contrast, matter is neither infinite nor eternal, since it was created and therefore there must be (More holds) a distinction between matter and space though this is denied by Descartes. How could matter move unless there were were empty space for it to move in, More asks, less plausibly? Moreover, there must be space empty of matter that is full of spirit, an extended though not an impenetrable substance for which room must be found in the universe. If the universe were full of matter to the exclusion of spirit, as Descartes supposed, natural philosophy would be nothing but materialism. More also argued that Descartes's identification of matter, extension and space might serve as a ground for the denial of divine creation. Modern critics regard this fear as well founded: 'so long as [Descartes] maintained that extension was the essence of matter (not of minds) and that matter was a substance, his teaching could not be rid of the implication that matter was uncreated and also that God and minds are nowhere' (Leyden 1968: 232). That is, spirit has no place in the physical universe.

I do not find More's concepts of space and time nearly as clear and precise as those to which Newton later gave expression. Newton established (for classical physics) that absolute space and absolute time were the true coordinates against which an absolute motion is to be measured; More has a conviction that motion can be absolute as well as relative, for 'the Cartesian definition of motion is repugnant to all the faculties of the soul, the sense, the imagination and the reason' (Koyré 1957: 144, quoting *Enchiridion metaphysicum*: 55) but he has little success in conveying a notion of the coordinate system against which motion should properly be measured; he calls it the 'inner *locus* [place]'. From the point of view of the *natural* philosopher it helps little to identify this with God. More succeeds better in defining the infinity of space than the infinity of time – which in his opinion, of course, can be associated only with God and with space, and not at all with the universe existing within time, which God can create and destroy and re-create, as More is fond of pointing out in his arguments. We might therefore distinguish *relative* time as Newton does (this is the time that is measured by astronomers or clocks), *universal* time (the finite duration of any universe) and *absolute* time which is the duration of God's eternity. More himself makes no such distinctions, however.

More's conceptions in opposition to Descartes seem, in effect, correlatives of his ideas of God's infinity and ubiquity. They are theological notions given a metaphysical dress, but certainly never formulated (as were the metaphysical presuppositions of Descartes, Leibniz and Newton) as foundations for a natural-philosophical system. Worthington's injunction to More (p. 158) was never to be obeyed.

More was both more careless in his language about space, and more bold in deifying it, than was Newton. The latter did not go beyond the assertion that 'by existing always and everywhere, [God] constitutes duration and space. Since every particle of space is *always*, and every indivisible moment of duration is *everywhere*, certainly the Maker and Lord of all things cannot be *never* and *nowhere*' (Cajori 1946: 545). Thus, like More, Newton refuted 'Nullibism' but he could hardly be supposed to mean that God constituted (that is, created or established) himself. Newton is careful to define what he understands by relative space in contradistinction from absolute space, which More does not do. In short: 'While More liked the thought of absolute space because it brought the spiritual so close to experience, Newton postulated absolute space in view of other needs. Unlike More . . . Newton considered it the subject of the highest importance in the system of the universe' (Power 1970: 290–1).

In contrast to the vague incompleteness of More's discussion, consider the clarity and definition of Barrow's treatment in his *Mathematical Lectures*. These were delivered at Cambridge in 1664 and subsequent years; it would be strange indeed if Newton had not been in the audience. The message of Barrow's first few lectures is that physical science is best treated mathematically, above all geometrically, and he explains why geometrical knowledge is certain. Barrow is a realist: the geometer is not concerned with imaginary or ideal lines or figures that exist only in his mind, though it is true that the magnitudes he is concerned with are perceived by the eye of reason rather than the eye of light: 'each and every geometrical shape that can be comprehended [*intelligi*] is really present in every particle of matter whatsoever; is present, I repeat, in actuality and in perfection, though it does not appear to the senses' (Whewell 1860, I: 84–5, trans, in Kirkby 1734; Burtt 1932: 144ff.). Thus mathematical physics is not only impeccable logically in its internal construction, but actually corresponds to the reality of Nature.

Mathematics and mathematical physics are concerned with magnitudes, and magnitudes occupy and measure out space (Whewell 1860, I: 149, Lectio X). What then is space, Barrow asks? Is it distinct from magnitude? To think of it as uncreated and eternal, so as to be independent of God, is abhorrent to both reason and piety. Some, like Descartes (and Aristotle before him) reason a way out of this difficulty by identifying space with extension, that is, material magnitude. But space is infinite while matter is not; there is space between worlds, and God can create new worlds still in the infinity of space. Moreover, motion necessarily takes place in space, each magnitude occupying its own portion; here Barrow differs from Descartes again. He concludes that space does really exist, distinct from magnitude, and is what exists when magnitude is absent. It is interponibility ('betweenness'), the potentiality of which magnitude is the actuality.

Despite a certain resemblance in their deductions of ideas of space from the infinity and ubiquity of God, Barrow and More have little enough in common. Notably, Barrow does not share More's belief in the infinity of material worlds, nor does he fill space with spirit. More argued along Aristotelian lines that spirit and space were related as substance and accident, that is, space inheres in spirit; he cannot accept space as pure dimension or potential dimension. Barrow rejected More's spirit, even though he seems reluctant to admit that any space *within* a world can be proved to be empty of matter. In other words (though there is some difficulty in following his discussion, as Barrow confessed to his auditors) he seems to conceive of a single geometric idea of space, which is common to all worlds, which exists where worlds are not, and which is the potential locus of all material existence. His space exactly anticipates Newton's conception of absolute space and is not theologically conceived in the manner of Henry More.

Just as space existed before the creation of worlds, so did time; just as space exists outside worlds, so does time. There is, declares Barrow, a great deal of analogy and affinity between space and time: 'For as space is to magnitude, so time appears to be to motion, so that time is, as it were, the "space" of motion.' And just as space is the potentiality of magnitude, so time is the potentiality of existence. The regular circular motions of the heavens are used by us to measure time, but absolute time has no dependence upon motion: 'the quantity of time is dependent neither upon motion nor upon rest; time flows with the same even tenour whether the Universe moves or is at rest, whether we sleep or wake' (Whewell 1860, I: 165; II:161). Suppose the stars to be at rest, the whole universe motionless and changeless, yet in this stillness of things time would continue. It is only in order to *measure* time that we must call on the aid of motion, choosing the most uniform and perfect that we can. For we can perceive nothing unless our senses are affected, and we know the measure of time only because our senses can perceive motion. We commonly take the celestial motions to be the most uniform accessible to us, and hence to be the primary measures of time, but this (Barrow says, somewhat mysteriously but prophetically) is not strictly true; the primary and original measures of time are rather

> those motions near to us which we observe with the senses and which are subject to our experiments, since by means of these we are able to judge of the regularity of the heavenly movements. Not even the Sun is the arbiter of time or a reliable witness, except in so far as his truthfulness is assured by mechanical time-keepers. (1860, II: 164)[5]

And Barrow adds that lacking such testimony we cannot be sure that the days of the past or the years of Methuselah were the same in absolute measure as our days and years.

Newton's mature views on these matters were given in the Scholium following the Definitions at the opening of his *Mathematical Principles of Natural Philosophy* (1687) (Cajori 1946: 6–12).[6] Like Barrow, he declares absolute time to be a uniform flow, or duration, without relation to anything external; relative time is what is measured by means of motion. We can approximate to absolute time by correcting solar time (as Barrow had indicated) by the 'equation of time', but we cannot be sure that we shall ever have access to 'an equable motion, whereby time may be accurately measured', in such a way as to be able to attain to absolute time exactly:

> All motions may be accelerated and retarded, but the flowing of absolute time is not liable to any change. The duration or perseverance of the existence of things remains the same, whether the motions are swift or slow, or none at all; and therefore this duration ought to be distinguished from what are only sensible measures thereof [that is, relative time]. (Cajori 1946: 8)

Similarly absolute space depends upon nothing external and is always and everywhere similar and motionless. And as the order of the instants of time is unchanging, so is the order of the parts of absolute space. Space and time cannot, as it were, move within themselves. Newton means that just as the instant in the succession of absolute time is unique, so is any place defined by the coordinates of absolute space, and only a movement with respect to such places is an absolute movement.

Just as relative time is measured by motion, so we measure out relative space with yardsticks, and for this purpose define a relative space with respect, for example, to the centre of the Earth. But if the Earth moves in absolute space,

> any space of our air, which relatively and with respect to the Earth remains always the same, will at one time be part of the absolute space into which the atmosphere passes, at another time it will be another part of the same, and so, absolutely understood, it will be continually changed. (1946: 6)

Hence, 'Absolute and relative space are the same in shape and magnitude, but they do not always remain numerically the same.' For what is constant in relative coordinates is changing in absolute coordinates, and vice versa. Note here again, as in Barrow, the highly geometric nature of the discussion.

An absolute place is defined by absolute spatial coordinates, a relative place by any coordinate system we choose to adopt. The absolute motion of a body is defined by the vectorial sum of its relative motions. Thus, to repeat Newton's example, a seaman moves relatively to his ship, which sails in relative motion to the Earth, while the Earth may be presumed to possess an absolute motion about the Sun. From these considerations the absolute motion of the seaman might be determined, but only (Newton reminds us) if we could be certain that the Sun and stars are absolutely at rest, which we cannot be in fact.

Therefore, since we have no confident knowledge of either absolute space or absolute time, we are forced to employ only relative measures of these quantities. However, Newton points out for the very first time, though we cannot always carry out the calculus for computing an absolute motion, we can surely point to some instances of it. Sometimes, it is true, whole systems of bodies move uniformly or are uniformly accelerated together in a straight line, so that it may be impossible for an observer either inside the system or outside it to know whether he is in motion or at rest, or whether either motion or rest is absolute, for 'any relative motion may be changed when the absolute motion remains unaltered, and the relative may be preserved when the absolute suffers some change' (Cajori 1946: 10). But it is otherwise with true rotary motion, for then there is a force of recession from the axis of rotation. If by a suitable experimental arrangement such a force can be made manifest, then it is certain that an absolute motion of rotation or revolution is taking place; the force would not be detected in the case of circular relative motion (such as the apparent revolution of the stars about the Earth). For example, we discover that the planets have a genuine force of recession from the Sun, and deduce confidently that they do really revolve about it in a true and absolute sense. 'It is indeed a matter of great difficulty to discover, and to distinguish effectually, the true motions of particular bodies from the apparent; because the parts of that immovable space, in which those motions are performed, do not by any means come under the observation of our senses' (1946: 12). And in fact the absoluteness of rectilinear motions can never be certainly determined, though (in practice) we may assume the Sun and fixed stars to be motionless.

With the exception of this last point – the question of the ontological distinction of circular from rectilinear motions – everything that Newton writes in the Scholium is a clear expression in terse language of what Barrow had declared more diffusely to his Cambridge audience twenty years before. Like Barrow's, Newton's reasoning is commonsensical and atheological, some of it springing straight from the kinematic tradition of Galileo. Like Barrow, Newton is endeavouring to extend the presuppositions of pure geometry, to move as it were from the arbitrary three dimensions of a mathematician's study to the permanent dimensions of space extending through the continuum of time. To me, Barrow's treatment and Newton's discussion in the Scholium read as close companion-pieces, and we can scarcely doubt the influence of the former upon the latter. More's opinions, however, though parallel in certain respects, belong to a different realm of thought and argument.

If, in search of further light upon the course of Newton's ideas, we turn back again from the printed Scholium of 1687 to Newton's private preparation for writing it, we perceive that *De gravitatione et aequipondio fluidorum* was much closer to Henry More. Like him, and for similar reasons, Newton criticized Descartes's definition of motion as translation, that is, the displacement of a body from one environment to another; in Descartes's view a body fixed in its environment is at rest, though the environment moves. Hence motion is to be thought of only in relative terms; this Newton in *De gravitatione* thought 'absurd'. If Descartes were right 'not even God himself could define the past position of any moving body accurately and geometrically now that a fresh state of things prevails, since in fact, due to the changed positions of the bodies, the place does not exist in Nature any longer' (Hall and Hall 1962: 130).[7] Therefore the possibility of a geometrical account of motion demands the existence of an absolute, unchanging system of coordinates, against which not relative but true or absolute motions can be measured. 'To avoid the Scylla of relativity, Newton embraced the Charybdis of absolute space' (Westfall 1971: 339). His choice was justified by a transcendental concept of absolute space and time emanating from God. In *De gravitatione* the concept of absolute space was made to depend more heavily upon the infinity of God than it was to do in the later Scholium. Extension – that is, space – Newton wrote, 'is not substance; on the one hand because it is not absolute in itself, but is as it were an emanent effect of God, or a disposition of all being', and on the other hand because space does not affect substance [matter] or minds (Hall and Hall 1962: 132). Nor is it an accident inherent in some subject, and it is certainly not nothingness. Space is infinite and motionless. It is a 'disposition of being *qua* being':

> No being exists or can exist which is not related to space in some way. God is everywhere, created minds are somewhere, and body is in the space that it occupies... space is an effect arising from the first existence of being, because when anything is postulated, space is postulated. (1962: 136)

Equally, Newton goes on, the existence of being postulates the existence of time, so 'the quantity of the existence of God is eternal, in relation to duration, and infinite in relation to the space in which he is present ... If ever space had not existed, God at that time would have been nowhere.' Created things occupy only limited regions of space and time. Space is indifferent to the presence of matter: 'space is no more space where the world is, than where no world is', by which Newton means that God's creation of matter did not also create space: God and space antedated the creation (1962: 137–8). There is, therefore, a differentiation between space and the void, or nothingness, and here Newton by implication approached tentatively to More's identification of space with God. As we shall see in the next chapter, he was more explicit in the even stranger metaphysical proposal that God is matter.

All this reads much like a transmuted echo of More, and the discussion is carried on with all More's acerbity against Descartes. But there is a strong recollection of Barrow also, in that Newton sought to geometrize space, though again perhaps the flavour is rather Platonic than Euclidean. Newton wrote that space is everywhere divided by surfaces infinite in extent and number, which also constitute boundaries, of all kinds and in all positions:

> hence there are everywhere all kinds of figures, everywhere spheres, cubes, triangles, straight lines, everywhere circular, elliptical, parabolical and all other kinds of figures, and those of all shapes and sizes, even though they are not disclosed to sight. For the material delineation of any figure is not a new production of that figure with respect to space, but only a corporeal representation of it, so that what was formerly insensible in space now appears to the senses to exist. For thus we believe all those spaces to be spherical through which any sphere ever passes, being progressively moved from moment to moment, even though a sensible trace of the sphere no longer remains there. We firmly believe that the space was spherical before the sphere occupied it, so that it could contain the sphere; and hence that as there are everywhere spaces that can adequately contain any material sphere, it is clear that space is everywhere spherical. And so of other figures. (1962: 133)[8]

This bizarre notion is expressed more tersely in the Scholium.

It is evident that Newton's thoughts about space and time underwent a change between the writing of *De gravitatione* and the composition of the *Principia* Scholium, a change taking Newton away from More and closer to Barrow. We have no way of accounting for this change or even dating it: to say that the writer of the Scholium was less willing to be drawn into theological metaphysics than the Newton of *De gravitatione* explains little. I think we may perhaps associate the change with Newton's increasing confidence, during the early 1680s, in the combination of mathematical analysis with mechanical principles (involving his own new conception of forces in Nature). This confidence was of course strengthened by his recent success in dealing with the problems of motion treated in the pre-*Principia* drafts of late 1684 and early 1685. His experience of philosophical controversy in the 1670s had taught him that it was *preferable* to avoid metaphysics altogether; his subsequent transformation of the mechanical philosophy persuaded him that it was *possible* at least to avoid exposing himself in theological metaphysics, in such a way as would lead to his identification with the Cambridge Platonic school. It was after all his business in the *Principia* to speak as a geometer, exactly as Barrow had done twenty years before in his mathematical lectures.

Or so Newton seems to have reasoned in 1687. In the long years of preparation for the second edition of the *Principia* (1713) Newton expressed his thoughts with greater freedom. The Scholium on space and time remained essentially the same from first to last, but he now added a theological conclusion to the whole work in the form of a General Scholium (*Scholium Generale* 1713, my translation; = Cajori 1946: 543ff.).[9] Here Newton plainly asserts that the complex motions of the heavens cannot arise from mechanical causes; the whole perfection of the system of the world could only spring from the designs and authority of a being both intelligent and powerful, who rules all things not as a 'soul of the world' but as the master of the universe: 'He is not eternity and infinity, but he is eternal and infinite. He is not duration and space, but he endures and he is present. He endures for ever, and is present everywhere and he constitutes duration and space by existing always and everywhere' (cf. Cajori 1946: 545). Thus, like More, Newton thinks that time and space are not themselves God but they are of God. Then Newton again echoes More in confuting the nullibists: 'As every particle of space exists for ever, and every indivisible moment of time exists everywhere, surely the creator and lord of all things cannot be absent from time and space.'[10]

We may by analogy, says Newton, affirm that as a man's soul pervades his body, so God pervades His universe:

> God is one and the same God always and everywhere. He is omnipresent not as in *virtue* only, but as in *substance*, for virtue cannot exist without substance. In him the universe is contained and moved, but without mutual interaction, for just as God has no feeling of the motions of bodies, so bodies feel no resistance from the omnipresence of God. (cf. Cajori 1946: 545)

How God is able so to permeate and govern the universe we can no more tell than a blind man can understand the nature of colours, or a sighted man see the inner structure of matter, but (Newton affirms) the relation of God to His universe is not that of man's soul to his body, which demands the intermediary of a brain and nervous system. We have no knowledge of the divine substance:

> We know God only by his properties and attributes, and his all-wise and excellent composition of things, and his final cause, and we wonder at his perfections; but we venerate and adore him because of his mastery of things ... From blind metaphysical necessity which is the same always and everywhere no variety of things would arise. All the diversity of creation, varying with time and place, could have arisen only from the ideas and will of a Being necessarily existing. (Cf. Cajori 1946: 546)

Again in the tradition of Henry More, Newton closes this passage with the bold declaration that 'to discourse of God from the phenomena [of Nature] does surely pertain to natural philosophy'.

Commentators have not failed to remark that in the revision of the *Principia*, Newton 'added a theology to his scientific system' that he had not felt needful in 1687. A theology was intellectually necessary because Newton knew his physical system to be incomplete, in that he could not account for its origins nor guarantee its perpetuity, and because only by appeal to theology could Newton's metaphysical foundations be justified; W. von Leyden writes:

> God's attributes constitute a spatial and temporal framework for the laws of mechanics and for what [Newton] accepted as the invariable, absolute characteristics of motion. Suppose that Newton had refrained from theo-

logical speculation concerning absolute space and duration: could he in that case have proved the existence of absolute motion, space and time solely on the basis of his principles of mechanics?

In my view the answer to this question is 'no'.

Without God as a guarantor of the absolute, Newton could not have safely left the world of relative motions (though, one might imagine, he could have postulated the absolute dimensions as axioms?). Accordingly, von Leyden holds that 'Newton's system is heterogeneous to the extent that it combines mathematical demonstration with metaphysical assumptions' (1968: 252).

Naturally enough, Burtt held a similar opinion, writing that

> Newton's doctrine is a most interesting and historically important transitional stage between the miraculous providentialism of earlier religious philosophy and the later tendency to identify the Deity with the sheer fact of rational order and harmony. God is still providence, but the main exercise of his miraculous power is just to maintain the exact mathematical regularity in the system of the world without which its intelligibility and beauty would disappear. (1932: 296–7)

I do not believe that the passage from natural philosophy to metaphysics and from metaphysics to theology seemed in any way improper or undesirable to Newton. *Pace* Thomas Hobbes (and perhaps Bacon) seventeenth-century natural philosophers were not eager to adopt a purely positivist or empiricist line, however appropriate such a line might seem in the day-to-day business of detailed scientific investigation. Descartes had founded his philosophy upon metaphysics and confirmed his metaphysics by the truth of God's existence. Newton resembled the vast majority of his contemporaries in possessing a firm belief in the reality of God the Creator, and he would have been surprised and mistrustful had his system proved so perfect as to exclude God. Von Leyden's argument could be inverted: for it might be contended that Newton insisted upon the doctrine of absolute dimensions – which are of no practical value in scientific work – precisely in order to demonstrate the perfect agreement of his scientific system with theology. Where he differed from Descartes and Leibniz was in making the divine essence continually active in time and space, and it was in this respect that the influence of Henry More may be seen as most powerful.[11] One might also argue that Newton himself can hardly have regarded his theological commitment as either a weakness or a superfluity in his scientific system, since he added it as a deliberate afterthought. I agree with von Leyden that it would be absurd to suppose that Newton was consciously deserting rational canons of scientific discourse when he drew discussion of God into natural philosophy – for he resolutely refuted Leibniz's accusations to this effect, that he made science a miracle.

Where does all this leave the question of Henry More's influence upon Newton? The majority of philosophical historians (Burtt, Koyré, von Leyden) argue two points:

1   The Scholium on space and time is concerned with metaphysical issues, upon which Newton's conclusions are crucial for his science.

In Koyré's words: 'Newton's physics, or, it would be better to say, Newton's natural philosophy, stands or falls with the concepts of absolute time and absolute space ...'

2   Newton's concepts of space and time are identical with those of Henry More, and may be presumed to derive from him.

To complete Koyré's quoted sentence: Newton's concepts of absolute time and absolute space are 'the selfsame concepts for which Henry More fought his long-drawn-out and relentless battle against Descartes' (1957: 160).

The powerful dissenting voice, in a persuasive article (1959), is that of Stephen Toulmin, who maintains that Newton in the Scholium lays down definitions required for a dynamical exposition, not for creating a meta-physical system. He emphasizes the point that for Newton these absolutes are mathematical; they are required for an axiomatic structure rather than a picture of reality; to ask whether absolute space really exists is absurd. The reader of the *Principia* is not compelled to tie Newtonian dynamics to the theological superstructure added to the concept of absolute space by Henry More and Joseph Raphson: 'The distinction between absolute, mathematical and relative, sensible space, time, and motion can be inter-preted consistently as a logical rather than a metaphysical distinction, and the theory as a whole justified on purely dynamical grounds.' Further, the thread of Newton's personal intellectual development, or (in other words) the historical origins of Newton's ideas, 'however suggestive or sympathetic he may have found Henry More's theological bizzareries', do not entitle us to pronounce upon the validity of Newton's scientific system or to assert that ' "Newton's physics … stands or falls with" the substantial or spiritual Space of the seventeenth century Neoplatonist theologians.' Toulmin also notes that Newton's language nowhere identifies God and space so strongly as does that of More, and that Barrow might with greater plausibility be regarded as More's antecedent (1959: 7; 214).

These cautionary words were necessary, for it has been too easy to place Newton the metaphysician and magician in the limelight, to the eclipse of Newton the mathematician and empiricist. But there are weaknesses in Toulmin's arguments. We may consider Newton as the author of a system of rational and celestial mechanics justified by its inner coherence and its conformity with experience. This is a system still useful to present-day scientists. But Newton did not present his system in this austere guise and (for various reasons) tended to strengthen rather than weaken its links with metaphysics and theology in his writings after 1687, as Toulmin admits. This does not indicate any discomfort on Newton's part with the necessity for such links. Further, those scholars who make the claim that the absolute concepts are axioms necessary for dynamics (or were thought to be so by Newton) are in danger of both eating and conserving their cake if they also argue that discussion of time and space, metaphysical discussion, is not an essential foundation for the sciences of mechanics and cosmology. Furthermore, reflections of this kind have no bearing at all on the question of interest to many historians: what was the course of Newton's intellectual development before the *Principia*? or on the question of interest to many philosophers: in what kind of intellectual context did Newton articulate for himself the mathematical principles of his book? There is a sense in which Toulmin may merely be affirming that Newton was wrong to involve God in natural philosophy. His article was written in 1959, when the historical evidence bearing on Newton's study of More, and on his composition of early metaphysical sketches, was not readily available. It would seem rash, now, to deny that Newton's mind from first to last was strongly inclined to a religious view of Nature, and that he had no wish either to write the natural philosophy of an unreal world, or to write the natural philosophy of the real world in such a way that the deepest questions about existence were to be excluded from its compass.

The root of Toulmin's contention seems to be the idea that despite the formal likeness between More's and Newton's ideas of space and time, the former are irrelevant to the latter because the purposes of the authors are different: More was finding his way to knowledge of God, Newton to knowledge of Nature. Von Leyden was surely right to claim, in a rebuttal of Toulmin's article, that such a separation of God and Nature is not to be found in Newton or Newton's age: Newton was conscious – like all philosophers of his time – that reliance upon reason and science alone left his system of the universe in an intellectual vacuum (1968: 259–60). How could he justify his ontology? To fill the void he was forced to have recourse to some variety of natural theology, of which More's appealed to him best; and this was also best equipped to cope with the enigma of force in Nature. In Newton's time – though not in ours – the filling of the ontological void might well be seen as bringing strength and plausibility to Newton's system.

Von Leyden adds a more technical point. Both More and Newton made some approach to the idea of time as a dimension. The long tradition of philosophy had been to regard time as something quite distinct in kind from space, and also to regard motion as a change in space. Galileo had with great originality perceived the 'supreme affinity between time and motion', so that with him time became a dimension in mechanics. Similarly, Newton's greatest mathematical innovation was to think of changing mathematical quantities as flowing in time. Concurrently, though perhaps coincidentally, the idea of the evolution of the Earth itself, through time, began dimly to appear in the late seventeenth century. Thus in several intellectual contexts people began to grasp the idea that the *rate of change* of events is significant, and that 'place' cannot simply be considered as an invariant concept independent of time. Newton, for instance, understood that for dynamical reasons no place in the universe can ever be twice the same. As von Leyden puts it:

> Both Newton and More, as also Gassendi, accorded space and time the same sort of status; that is, they treated these concepts conjointly and analogously, regardless of whether their method of approach was logical, physical or metaphysical ... It might be said that part of the philosophical significance of both More's and Newton's doctrines is that they provided one preliminary

step towards the modern conception of a compound of space and time, the
'space-time' of modern physics. (1968: 259–60)

Perhaps this last suggestion may seem a little far-fetched. However, von
Leyden justly indicates – and Toulmin might applaud – a richness of content
in Newton's scientific thought which is independent of theology. Certainly
Newton did not simply borrow ideas from either More or Barrow. He
transformed them, and found new uses for them.

While Toulmin has argued that the absolute concepts of time and space
are axiomatically necessary as tidy foundations for dynamics, Burtt before
him had asked why they need be invoked at all. He distinguishes Henry
More's discussion of space and time from that of Barrow – and, for that
matter, from the views of the mature Newton – because Barrow's treatment,
though it shared More's fundamental metaphysical assumption that we are
assured of the homogeneity, permanence and eternity of space and time by
the existence of an unchanging, perfect deity (and at root this is the
Cartesian position also), in addition develops a purely scientific, or one
might better say geometrical, handling of these questions. At this point
Burtt asks the question why Barrow 'does not drop the absolutistic terminology and treat space and time as purely relative to magnitude and
motion, inasmuch as practically that is how they must always be treated?'
(1932: 152) Burtt might equally well have posed the question of Newton's
famous Scholium.

The answer to this question, tantamount to asking why Barrow and
Newton were not positivists, has been given partly by von Leyden, partly
by Toulmin, partly by Burtt himself in discussing Newton. For (unlike More)
in the discussion of space and time in the Scholium, Newton makes no
mention of God. His argument in favour of absolute dimensions is aprioristic,
experiential and commonsensical. It takes the step that Barrow did not
completely take of abandoning More's theological considerations altogether. But Newton does not end on a positivistic note. Now it may be
argued that if we juxtapose with the Scholium other, later writings of
Newton we may satisfy ourselves that Newton really did believe in God's
existence as a justification for the absoluteness of space and time, though
he excluded this belief from the Scholium. This argument leaves something
of More in Newton's mind. But if we consider the Scholium alone, as
Newton printed it in 1687, when the reader could not be cognizant of

Newton's 'private' thoughts about God and Nature which were only to be published afterwards (if at all), and therefore consider the Scholium honestly and not as a kind of subterfuge, we can see another philosophical, indeed mathematical, basis for the absolute dimensions of Barrow and Newton. It is Euclidean rather than Platonic idealism. Is not the absolute concept of space or time analogous to Euclid's 'absolute' notions of a straight line or a circle? And is not Burtt's wonder that Barrow (or Newton) did not become a positivist rather like asking why the geometer does not define a straight line in terms of a ruler, and a circle in terms of compasses – or indeed any concept whatever in instrumental terms? It seems to me that if Nature is to be geometrized – modelled in terms of Euclidean idealism – then it is only homogeneous to treat space and time universally as other dimensions are treated universally, that is as infinite absolutes. Mathematical physicists idealize absolute space and absolute time as constituting the coordinate framework of their science by analogy with the geometer's idealization of the straight line and circle from the figures of experience.

This major step, so far beyond More, was taken by Newton.

11

More and Newton: Force

A PHILOSOPHICAL minefield which the historian of seven-teenth-century scientific ideas cannot wholly avoid is the origin of Newton's concept of force; not only the concept itself, but the ways in which he employed it (both mathematically and rhetorically) to transform scientific theory. The concept of force, or more generally 'activity', is the thorniest point in Newton's philosophy of Nature, and was that which gave contemporaries greatest difficulty. Like space and time, force could not be taken up with the fingers, or in a beaker; as a cause it could never be isolated from its effect, motion, and vice versa (certain mathematical asymmetries in this context were not to be sorted out in Newton's lifetime). Force could be quantified, as the product of mass and acceleration, and classified qualitatively as gravitational, magnetic, chemical, etc., but it could not be homogenized – not all forces seemed to obey the same rules. Did it really exist, therefore, or was it just a name mistakenly applied to phenomena of different natures? If it existed at all, where did it come from?

Force in Newton's philosophy has a variety of different meanings and applications. In the first place, it can be both attractive and repulsive, as in magnetism; Newton himself was the discoverer of electrical repulsion. Again, it can be either long-range or short-range, with a very important distinction between the two classes. Long-range forces originate between the fundamental particles of bodies, as do short-range forces, but they also operate between macroscopic bodies to the largest size, which the short-range forces do not. Therefore the long-range forces (gravity, magnetism, electricity) cause distinct and perceptible motions in bodies, in the latter two instances, however, only certain kinds of bodies. About the phenomena produced by these forces there could be little doubt – though Newton was to introduce a new class of gravitational phenomena by extending gravity to the heavens. In each instance the phenomena of motion were clearly visible.

In the *Principia*, Newton was predominantly concerned with the long-range force of gravity, and particularly again with the mathematical deductive system which can be built upon the knowledge that this force follows a particular law, the inverse-square law. Newton also tells his reader that such an operating force can be measured by determining the mass and acceleration of the body moved by it. But to be able to treat quantitatively the manner of operation of such a force, as Newton does, tells us nothing about the origin of force in general and about its relation to matter. Newton's conspicuous success in quantifying dynamics created further problems about the nature of energy (to use our term) in the universe which Descartes appeared to have avoided; with Newtonian dynamics the mechanical philosophy became, as it were, open-ended, no longer the closed loop established by Descartes when he declared that the quantity of motion in the universe is constant.

Although Henry More knew nothing of Newton's force, he may have helped to mould Newton's ideas. The modern mechanical philosophy most carefully examined by More was that of Descartes, and this he rejected not only on broad metaphysical and religious grounds but because he could not understand Descartes's purely relative idea of motion, nor Descartes's treatment of motion and rest (see chapter 8). Therefore he did not accept Descartes's closed world of motion nor his idea of the indifference of matter to the states of motion and rest.

More's implicit and unquestioned assumption throughout his debate with Descartes of the absoluteness of true motion led him to assert that true rest is a real privation marked by a *renixus*, or inertia, a reluctance to be moved in the motionless body (Descartes 1974: 380). Correspondingly, motion is a force or action forcibly separating two bodies and capable of overcoming their resistance to motion. To Descartes this resistance was a mere prejudice founded upon our physiological experience of the difficulty of pushing and pulling heavy bodies. There is (he knew) as great a difficulty in stopping them when once they are in motion; this indeed is Cartesian mechanical force, the 'force of a body's motion'. Nevertheless, Descartes's treatment is essentially kinematic while More, his critic, initiates a trend towards a dynamic analysis of motion.[1] Descartes perhaps puts the point more clearly than does More himself: 'you conceive that there is a certain force in a body at rest, rendering it resistant to motion, as though that force were a positive something, or indeed a kind of activity, distinct from absence of motion, when it is plainly nothing at all distinct from a modal entity' (1974: 404; cf. Westfall 1971: 69). Though neither Descartes nor More developed the implication, it is clearly implied in More's argument that a propulsive force is needed to surmount the force of resistance. Furthermore, this propulsive force cannot be simply a mechanical transfer of the motion already present in the moving body that strikes the body at rest, for More cannot conceive such a transfer to be possible:

> Lastly [he wrote in his third letter to Descartes] I am quite bewildered when I think that so trifling and humble a thing as motion, separable from its subject and shifting about, and of so feeble and evanescent a nature that it would disappear at once unless it were reinforced by its subject, yet is able so strongly to master its subject and drive it forcibly about this way and that. Indeed, I incline rather to the opinion that there is actually no transfer of motion at all, but that one body is as it were awakened into motion by the impact of another, as the mind is aroused to thinking by this or that circumstance; and that a body does not so much receive motion [from another] as put itself in motion, when instructed by another body. And that, as I said just now, motion is to a body as thought is to the mind, that is to say, neither is received [from without], but both arise from the subject in which they are found. (Descartes 1974: 382–3)

More's comparison is surprising enough already: that motion is the product of a material body, as thought is of the human mind. But note that he does not say that *matter* is self-moving, and in the next sentence he goes on – more boldly still – to explain why: 'And that all we call body lives in a dull and sottish way, so that it is the last and lowest image and shadow of the divine essence, which I hold to be the perfection of life; though body is devoid of sensation and consciousness' (1974: 382–3, 23/7/1649, my translation; cf. Gabbey 1982, 211).[2] So there is a rudimentary form of life, which is to say spirit, in the lump of clay, as there must be still more of spirit in the active magnet. It is spirit that moves, as it is spirit that causes the rebound of elasticity, the fall of a stone and the attraction of lodestone or amber.

We have no evidence that Newton read the above passage, none that the printed versions of More's letters stood in his library. If he had read it, it might have appealed to him. In *De gravitatione* Newton makes the distinction between matter and space the immediate and continual result of God's will; it can hardly have been foreign to the same mind that motion, or rather one should say all the activity of the universe, should likewise be the immediate and continual result of God's will. Some of Newton's later utterances are quite consistent with such a view of things. However that may be, and we need not pursue the conjecture here, we may also note that Newton, like More, in his mature years makes the inertia of matter a 'positive something' and indeed a force, the *vis insita* puzzling to modern commentators: 'The *vis insita* of matter is a power of resisting by which every body, as much as in it lies [*quantum in se est*], continues in its state, whether it be of rest, or of moving uniformly in a straight line' (1687: 2, Definition III, my translation; see also Cohen 1964). In explanation of this Definition III in the *Principia* Newton adds that this *vis insita* 'does not differ in any way from the inertia of mass, save in our manner of conceiving of it'. Inertia causes a body to change its state only with difficulty, therefore the *vis insita* can very significantly also be called *vis inertiae*. Furthermore, 'a body only exerts this force when another force, impressed upon it, endeavours to change its state [of motion or rest].' This sounds almost like the behaviour of More's body 'as it were awakened into motion by the impact of another' and surely More would have approved of Newton's definition of inertia as a true 'power of resisting' changes in the state of motion. Inertia, as conceived by Newton, certainly 'arises from the subject in which it is found'. Another sentence in the passage just cited also concurs with More's notion of resistance or inertia as the counterpart of active power or force, for Newton writes that the 'exercise of this force is under different respects, as both resistance and impetus [momentum] ... it is resistance so far as the body, in order to maintain its state, struggles against the force of the impressed body; it is impetus so far as the same body, yielding with difficulty to the force of resistance of the obstacle, endeavours to change the latter's state.'

The animistic language of 'struggle' (*reluctatur*) and 'yielding with diffi-culty ... endeavours' (*difficulter cedendo, conatur*) would surely have appealed to More, who would have contended that something in these bodies must apprehend what is happening to them, before reacting appropriately. He would have said that Newton was attributing to matter capabilities different in kind from such invariant, passive ones as hardness and impenetrability, qualities that could not arise from such mechanical properties alone (1687: 2, Definition III, my translation; cf. Hall and Hall 1962: 241).[3] All in all, it seems not unreasonable to claim that Newton's definition of this basic attribute of matter adheres (in its precise terms) less to Descartes's for-mulation in his Rules than to More's modification of Descartes. It may be that this reflects rather a similarity in the set of the ideas of the two philosophers from Lincolnshire than a direct textual influence.

It might be argued, perhaps more fancifully, that Newton was made aware by More of the necessity for *cause* in considering the change of the state of motion in bodies, a necessity which Descartes had not felt. More found the cause in his Spirit of Nature; Newton found the *agent* of change of motion in 'impressed force', leaving the ultimate cause and origin of force undetermined, In *De gravitatione* Newton had written:

> Force is the causal principle of motion and rest. And it is either an external one that generates or destroys or otherwise changes impressed motion in some body; or it is an internal principle by which existing motion or rest is conserved in a body, and by which any being endeavours to continue in its state and opposes resistance. (Hall and Hall 1962: 148)[4]

In the *Principia* he was less explicit about the status of force as a cause, but makes it clearer that impressed force only acts for so long as the change in motion continues. The action may be instantaneous, or continuous as with gravity and other centripetal forces, but it is always a cause of change of motion, that is, acceleration. Newton did not imagine, as Descartes did, that necessarily the unique force in Nature is that of impact; in Definition VI he speaks of the 'effectiveness of the cause that propagates centripetal force from a centre through the adjacent region'. This is very much like an enunciation of the idea of a field of force, but of course (as his correspondence with Richard Bentley makes plain) Newton could not explain how the 'field' is maintained in space or how it acts upon bodies. As he there wrote, he could only resign the question of whether the 'field' is material or immaterial to the consideration of his readers (Hesse 1961: 148–56).

The idea that the revolutions of the planets and the tides of the oceans might have mechanical causes, as old as Kepler and Galileo, was established for Newton's generation by Descartes, and hence the key metaphysical issue for the *Principia mathematica* was this one of 'action at a distance', or the mystery of the gravitational 'field' surrounding each particle of matter. When Newton came to extend the notion of force to the invisible realm of the constituent particles of matter, postulating the existence of optical, cohesive or chemical, electrical and perhaps other short-range forces, though not more mechanistic than Descartes had been in his hypotheses, he was very largely extending the conceptual bounds of the mechanical philosophy of Nature. No one before him had stretched the qualitative distinction between the phenomena of chemistry, optics and so on to a qualitative (and perhaps quantitative) distinction between the operating causes or their agents, that is forces, while preserving the conceptual unity of the idea of corpuscular displacement and rearrangement (already developed by Robert Boyle). So strange was this step from the macroscopic to microscopic, from long-range, experiential force to short-range, inferential force that Newton felt impelled to justify it in the third of the Rules of Reasoning added to the second edition of the *Principia*.[5] It is true that in this Rule – asserting that in making inferences about the fundamental structure and properties of matter from our experiments on gross bodies we must neither be misled by dreams nor fail to follow the 'analogy of Nature which is wont to be simple and ever consonant to herself' – Newton made a point that was essential to the construction of the theory of gravitation, yet one even more vital for any beginning upon the theory of short-range forces (1713: 357–8 = Cajori 1946: 398–9).

We have known for a quarter of a century that Newton at one time meant to treat in his *Principia mathematica* of these short-range forces, as well as the long-range force of gravity (Hall and Hall 1962: 302ff.; 320ff.). In the book ultimately composed there were given a few examples of how they might be treated by the geometer – for example, the propositions showing how optical phenomena might be caused by a short-range force in matter attracting the particles of light – but in the main Newton could only hint at the parallelism between the two kinds of force; gravity was to remain the unique force that he could subject to an extensive and profound mathematical analysis. Yet the dream of a vaster scope of things remained in the Preface to the book:

> Would that it were possible to derive the rest of the phenomena of Nature from mechanical principles by the same kind of argument! For many considerations lead me to suspect that all of them must depend upon certain forces by which the particles of bodies (in accordance with causes hitherto unrecognised) are either impelled towards one another and so cohere in regular patterns, or else are repelled and recede from each other. These forces being unknown, philosophers have as yet grappled vainly with Nature. (1687: Preface, my translation)

For the complete range of mathematical philosophy the Rules of Reasoning were the more necessary because the short-range forces were involved in the production of motions that are normally invisible, motions of the fundamental particles (or larger, compound corpuscles) inferred as occurring in chemical reactions, or in interactions between bodies and light (such as refraction and reflection); or again the force of cohesion, cognate to inertia, which does not so much produce motion as inhibit it by resisting disruptive forces. Only the short-range force associated with capillary action, carefully studied by Newton and Hauksbee – if this force be distinct from the force of cohesion – produced a visible movement of gross matter. Since the particulate motions whose existence was inferred from the phenomena by Newton, as by Boyle and other philosophers before him, were undetectable directly, the forces causing them were a stage more remote even than the forces of gravity or magnetism. In the *Queries* appended to *Opticks* Newton was to give many examples of the action of the short-range chemical force:

> when Salt of Tartar runs *per Deliquium*, is not this done by an Attraction between the Particles of the Salt of Tartar, and the Particles of the Water which float in the Air in the form of Vapours? And why does not common Salt, or Salt-petre, or Vitriol, run *per Deliquium*, but for want of such an Attraction? ... And when Water and Oil of Vitriol poured successively into the same Vessel grow very hot in the mixing, does not this Heat argue a great Motion in the Parts of the Liquors? (1952: 376–7)[6]

And Newton there noted the distinction between these short-range and the experiential long-range forces:

The Attractions of Gravity, Magnetism and Electricity reach to very sensible distances, and so have been observed by vulgar Eyes, and there may be others which reach to so small distances as hitherto escape Observation; and perhaps electrical Attraction may reach to such small distances, even without being excited by friction. (1952: 376)

In which case, obviously, electricity would be anomalous in being the sole short-range force that could be raised to the long-range level by experiment.

The unfolding of Newton's ideas about the activity of short-range forces had to wait for some twenty years after the *Principia* for a discursive and non-mathematical treatment in these *Queries*. The *locus classicus* is *Query 31*:

Have not the small Particles of Bodies certain Powers, Virtues, or Forces, by which they act at a distance, not only upon the Rays of Light for reflecting, refracting, and inflecting them, but also upon one another for producing a great Part of the Phaenomena of Nature? For it's well known, that Bodies act one upon another by the Attractions of Gravity, Magnetism, and Electricity: and these Instances show the Tenor and Course of Nature, and make it not improbable but that there may be more attractive powers than these. For Nature is very consonant and conformable to herself ...

... And thus Nature will be very conformable to herself and very simple, performing all the great Motions of the heavenly Bodies by the Attraction of Gravity which intercedes those Bodies, and almost all the small ones of their Particles by some other attractive and repelling Powers which intercede the Particles ...

... Seeing therefore the variety of Motion which we find in the World is always decreasing, there is a necessity of conserving and recruiting it by active Principles, such as are the Cause of Gravity ... and the Cause of Fermentation ... For we meet with very little Motion in the World, besides what is owing to these active Principles. (1952: 395–6; 397; 399)

Newton was as certain as Henry More that though these powers, virtues or forces are invariably associated with matter, they are not produced by matter, with the possible exception of the *vis insita*, or force of inertia, since it seemed impossible to separate the concept of matter from the concept of this force. (Fortunately inertia – if a force at all – seemed to be qualitatively distinct from the 'active' forces.) In consequence, Newton faced the same problem as More before him, and Descartes before More: the problem going far back in Greek philosophy: where does the activity of Nature come from? Matter makes the Universe exist in space and time: what makes it 'go'? A principal problem in seventeenth-century natural philosophy was to find an escape from the following dilemma: if matter is active, is it not then totally self-sufficient? But if matter is wholly passive, what is the active agent of its changes? Descartes had founded his science of Nature on the belief that perpetual motion is this agent; More had declared that it is spirit, or God. Neither view satisfied Newton who indeed sought, as far as he could, to banish the dilemma from mathematical physics (for which there were Cartesian precedents too). In the *Mathematical Principles* he took a positivist line: if we postulate the demonstrable fact of gravitational attraction proportional to the quantities of matter and the inverse square of the distance, we can then construct a mathematical theory that can be tested by observation and experiment. But many contemporaries found such a vast physical postulate impossible of acceptance. And in later life – as in the Queries added to *Opticks* – Newton speculated more and more openly, and confusingly, about the nature and mechanism of force or activity.

Our concern here is only with the possible relation of Henry More to this aspect of Newton's thinking, a question recently reopened by J. E. McGuire (1977). His account sets Newton with More in the context of Cambridge Platonism, in reaction to the suggestion that Newton 'borrowed' the idea of active principles in matter (or in association with matter) from the alchemists, an interpretation of their origin advanced by R. S. Westfall (1972; 1975) and B. J. T. Dobbs (1975: 210–13). In 1972 Westfall suggested that 'Newton's concept of forces between particles derived initially from the world of terrestrial phenomena, especially chemical reactions.' Newton had taken the 'animism and active principles of the Hermetic [or alchemical] tradition' and combined them with the ideas of the mechanical philosophy, thus leading him 'in the end to an expanded view of a mechanical universe through which concepts of Hermetic origin were permanently built into the structure of modern science' (1972: 190–1; 193). Newton's language of attraction and repulsion between particles of matter originated, in Westfall's view, within the 'Hermetic tradition' but the combination of these ideas with the mechanical philosophy was no betrayal: 'the active principles of the Hermetic tradition could lead the relatively crude mechanical philosophy of 17th century science to a higher plane of sophistication. The Hermetic elements in Newton's thought were not in the end antithetical to the scientific enterprise' (1972: 195).

While Westfall attributes Newton's mathematization of the active principles to a late date (the 1680s), as is unavoidable, he finds evidence of Newton's adoption of force as an active principle more than a decade earlier, in *De gravitatione*, quoting passages cited above (p. 228). Presumably at least by the 1670s, if not before, Newton had (in his view) begun to think of chemical and optical phenomena also as attributable to such principles. Finally he recognized that gravity could be considered as of the same kind, and so as being an attractive force.

Several objections to these hypotheses can be raised but it is not possible to doubt Westfall's assertion that Newton was well versed in chemical and alchemical authors by 1670, and was already conducting his own experimental investigations of their teachings. The argument that Newton was thinking in terms of short-range forces or active principles before his thoughts were turned to planetary motion by Robert Hooke in 1679 is less well founded. Passages indicating Newton's attachment to a vaguely animistic view of Nature, in which matter was seen as brute and inactive (as it was also by Newton's friend John Locke), can certainly be found but (as Westfall himself hints) these suggest a Platonic rather than an Hermetic influence. Newton's clear discussions of non-gravitational forces are all of post-*Principia* date; they are clearly based upon and also contrasted with his theory of gravitational attraction. Westfall's hypotheses ignore the possible connections between Newton's ideas of attractive force and other concepts of attraction current throughout the seventeenth century which are in no way indebted to Hermeticism. Kepler had borrowed William Gilbert's notions of magnetic attraction and magnetic polarity for use in the context of planetary motion. True, Newton did not read Kepler; but he did refer explicitly to Giovanni Alfonso Borelli's modification of Kepler's ideas. Newton pleaded ignorance of what Robert Hooke had previously published concerning 'the compounding the celestiall motions of the planetts of a direct motion by the tangent & an attractive motion towards the centrall body' but he did not deny that this form of words occurred in Hooke's letter to him of 24 November 1679. There is no need to suppose that Newton had to search in dark corners for the concept of attractive force, or indeed for the concept of an active principle or principles in Nature. There was a tradition of 'attractive force' and 'action at a distance' strong and lively in the seventeenth century, despite Cartesian opposition to it, which may have been related to Platonism (as with Henry More) or the ideas of Paracelsus but had little to do with alchemy and Hermeticism.

It is well known that Newton, in his anxiety to protect himself from neo-Cartesian criticism, was concerned to separate his invocation of attractive force as much from the attraction of Roberval's *Aristarchus* (1644) as from the action at a distance of Kenelm Digby's weapon-salve. Occult qualities were imagined (he wrote) 'to lie hid in Bodies, and to be the unknown Causes of manifest effects' (1952: 401). He was not proposing anything of that sort but a law of action amply confirmed by empirical investigation. The interpretation of Westfall and Dobbs would couple Newton firmly with the line of thought he overtly rejected – but then he also detached himself from More, whose influence upon Newton is (I suggest) so much more clear. I do agree with Westfall that Newton perceived a necessity for a source of activity in Nature that could not be found in inert matter – indeed, Newton said this many times. I also agree with Westfall that we may see More's example and teaching in this trend of his thought. However, I see no need to invoke any alchemical precedent for the ideas of attraction and repulsion, and of active force.

But it is time to return to the activity-generating forces or principles of which Newton wrote in the Preface to the *Principia* and the *Queries* in *Opticks*, most prominently, the cause of gravity. It is clear that in the few years between 1679 and 1684 (when the preparation for the *Principia* began) Newton finally renounced any faith in the Cartesian aether. Later, he would speculate about very different aethers, but these are really irrelevant to the question of More's possible influence upon Newton. Newton did not, it seems, reject the Cartesian aether for the same reasons of principle as More, but because of the scientific difficulties it created. The planetary vortices are an integral part of Descartes's system of physics; in the later 1670s, however, Newton learnt that if Kepler's laws of planetary motion are accepted the vortex is mathematically unnecessary, and in 1685 he discovered that the fluid vortex is actually inconsistent with Kepler's laws: the proof of this was given at the end of Book II of the *Mathematical Principles*. Newton's own private Ariadne's thread through all such thinking, leading to his final rejection of Cartesian mechanism, was geometry rather than the metaphysical and theological reasoning of Henry More.

If Descartes was mistaken, in Newton's opinion, in his postulation of aethers transmitting the perpetual motion of the universe hither and thither, so equally mistaken was that other version of the crude mechanical philosophy (represented by Boyle's being content to assume, as a first hypothesis, that air-particles are inherently springy) which attributed to the basic particles the properties of macroscopic bodies such as cohesion (Descartes's rest), or particular shapes (Descartes's screw-particles) or elasticity. Such ideas seemed to Newton as feeble as the hooks and eyes of the Greek atomists. '[The] vast Contraction and Expansion [of the atmosphere] seems unintelligible [Newton wrote], by feigning the Particles of Air to be springy and ramous, or rolled up like Hoops, or by any means other than a repulsive Power' (1952: 396).[7] As he pointed out in the third Rule of Reasoning of the *Mathematical Principles*, it was fair to infer from the macroscopic world that the essential nature of matter (possessing mass, being hard, being impenetrable) was the same in the microscopic world, but to give particles qualities properly belonging only to large compound bodies was simply to postpone, not to resolve, the problem of explaining them. Here once again — at least in relation to the 'monstrous' spring of the air — Newton accepts a criticism formerly advanced by More, but he does not also accept More's alternative, the Spirit of Nature or Hylarchic Principle, or the 'Plastick Nature' of Cudworth. In *De gravitatione* he envisages the idea that a 'soul of the world', created by God, might exercise divine functions in Nature mediately, like More's Spirit of Nature, but he sees no advantage in this idea over the notion of God's direct action (Hall and Hall 1962: 142).[8]

However, in the same unfinished essay, Newton devoted many pages to the exploration of an idea as idealistic and as theologistic as any expressed by Henry More himself. Like some other idealists and some mathematical physicists, Newton found the existence of brute matter (even as God's creation) uncomfortable: it entailed the awkward dualism of mind and matter. Therefore, he asks, could we not imagine that all material existence is simply the realization of a thought of God? Then mind and matter as created beings will be all one, for if God ceased to will the realization of his thought, matter would cease to be.

To give precision to this rather extravagant notion, the youthful Newton worked it out as follows:

> we may imagine that there are empty spaces scattered through the world, one of which, defined by certain limits, happens by divine power to be

> impervious to bodies, and *ex hypothesi* it is manifest that this would resist the motions of bodies and perhaps reflect them, and assume all the properties of a corporeal particle, except that it will be motionless. If we may further imagine that that impenetrability is now always maintained in the same part of space but can be transferred hither and thither according to certain laws, yet so that the amount and shape of the impenetrable space are not changed, there will be no property of body that this does not possess. (Hall and Hall 1962: 139)

These volumes of space upon which the divine will has conferred the properties of particles will be able to affect human minds, and will either be like bodies or actually *be* bodies: 'If they are bodies, then we can define bodies as *determined quantities of extension* [space] *which omnipresent God endows with certain conditions*' (1962: 140). What is the usefulness of this conception? It is that it gives us an enhanced view of the divine creativity:

> it clearly involves the chief truths of metaphysics, and thoroughly confirms and explains them. For we cannot postulate bodies of this kind without at the same time supposing that God exists, and has created bodies in empty space out of nothing, and that they are beings distinct from created minds, but able to combine with minds. (1962: 142)

Now created matter is neither eternal nor self-sufficient: it is dependent upon the constant and continuing will of God. Here Newton echoes More precisely:

> If we say with Descartes that extension is body, do we not manifestly offer a path to Atheism, both because extension is not created but has existed eternally, and because we have an absolute idea of it without any relationship to God, and so in some circumstances it would be possible for us to conceive of extension while imagining the non-existence of God? Nor is the distinction between mind and body in this Cartesian philosophy intelligible ... (1962: 142–3)

Here, it seems to me, Newton has made very clearly an extremely More-like point: we must conceive extension (space) to be coeval with God, and to precede, indeed to be the condition for, the creation of matter, otherwise we run the risk of committing the grievous error of making matter coeval with God.

This theory of the dependence of matter on the continuous exertion of the divine will was never made known to anyone by Newton, but a cognate idea, that the active principles or forces in Nature may be products of the continuously active will of God was acknowledged by Newton, if guardedly. Talking with David Gregory about his proposed additions to the Latin translation of *Opticks* (1706), Newton was thus recorded:

> His doubt was whether he should put the last Quaere thus. *What the space that is empty of body is filled with.* The plain truth is, that he believes God to be omnipresent in the literal sense; ... for he supposes that as God is present in space where there is no body, he is present in space where a body is also present. But if this way of proposing this his notion be too bold, he thinks of doing it thus. *What Cause did the Ancients assign of Gravity.* He believes that they reckoned God the Cause of it, nothing els, that is no body being the cause; since every body is heavy. (Hiscock 1937: 30, 21/12/1705)

Thus Newton meant to attribute his metaphysics not to Henry More, whose ideas he so precisely mirrored, but to the philosophers of antiquity; and perhaps More would not have disapproved of such a use of the perennial philosophy.

In the final version Newton printed a passage of lower key – nevertheless running into trouble with Leibniz who was well enough apprised of the true tenor of Newton's thought – which is little more than a statement of God's omnipresence in his creation, and of the ability of his skill and wisdom to move bodies within the universe so as both to form it and to reform it.[9]

Gregory's memorandum might be read as a gloss upon a famous passage in a letter to Richard Bentley of 1693 (printed sixty years later) where Newton wrote

> It is inconceivable, that inanimate brute Matter should, without the mediation of something else, which is not material, operate upon and affect other Matter without mutual Contact ... That Gravity should be innate, inherent and essential to Matter, so that one Body may act upon another at a Distance thro' a *Vacuum*, without the Mediation of any thing else, by and through which their Action and Force may be conveyed from one to another, is to

me so great an Absurdity, that I believe no Man who has in philosophical Matters a competent Faculty of thinking, can ever fall into it. Gravity must be caused by an Agent acting constantly under certain Laws; but whether this Agent be material or immaterial, I have left to the Consideration of my Readers. (Cohen 1958: 302–3)[10]

Bentley understood and shared Newton's concern to make science serve and support the cause of true religion, and he demonstrated at length and with vigour that the Newtonian system of the universe founded upon universal gravitation could only come into being, and continue, by God's creative power. He was perhaps less happy in rendering Newton's idea of God's mediation in the universe under the form of gravitation. Gravity, Bentley wrote, is not innate and essential to matter, and therefore it 'could never *supervene* to it, unless impressed and infused into it by an immaterial and divine Power' (1958: 341).[11] This blurs Newton's clear contention that the 'Agent' effecting gravitation must be *continually* at work; that even God cannot make matter, as such, unaided, attract other matter at a distance without something else – which of course, for all Newton's caution, cannot be material – filling the space between.

In all this Newton remembered the lessons that More had taught. It can surprise no reader that the more deeply Newton enters into metaphysics, the sharper his recollection of More becomes; and that he is most dependent on More in theological metaphysics.

This might have been, but is not, the end of the story. If it had been it would be possible, perhaps, to generalize with J. E. McGuire on the contribution made by Cambridge Platonism, and More in particular, to the Newtonian understanding of forces:

> In general, ... the Neoplatonism of active and passive principles, as they are developed by the Cambridge Platonists, could clearly legitimize the general notion of phenomena that act at a distance but not necessarily the specific, though relatable, Newtonian concepts of attractive and repulsive force. Given Newton's commitments, however, experience of the action of phenomena such as magnetism, electricity and chemistry, the operation of distance forces in planetary astronomy, and perhaps the activity of certain alchemical processes would almost certainly suggest the empirical characteristic of attraction and repulsion. Analysis could then provide the basic

parameters needed for their theoretical resolution with respect to the action of specific phenomena. (1977: 104)

While it would be a notable error to link Newton with the unmitigated animism and distrust of all secondary explanations that seemed to overwhelm Henry More in his last years, it would be hard to deny the permanent traces, perhaps one might even say the significant permanent influence, of the older Lincolnshire philosopher upon the younger one. And yet it must also be allowed that at least as phrased by its champion, McGuire, More's effect upon Newton seems to involve a kind of pompous nebulosity not typical of either man. Objections spring to mind: More expatiated on passive matter and active spirit, but I have found nothing in his writings of passive and active *principles* (or forces). Nor does he focus on the problem of action at a distance. The language of attraction and repulsion in relation to magnetism and electricity Newton would have found in quite different contexts, while the extension of this to light is all Newton's own. And so on. It is most plausible to see the ideas in the *Queries* as a complicated synthesis from many sources.

But Newton was to change his mind again and, as some would have it, renounce the explanation of force in terms of the omnipresent activity of God which he apparently held until at least 1706. Henry Guerlac has examined the effect of Francis Hauksbee's investigation of capillarity and especially of his dramatic electrostatic experiments upon Newton's subsequent pronouncements (1963b; 1967). They inspired Newton to add further *Queries* to the new edition of *Opticks* (dated 1717), those numbered 17 to 24 in the current text, and four extra pages to *Query* 31. He here introduced 'a much subtiler Medium than Air, which after the Air was drawn out remained in the *Vacuum* ... the same with that Medium by which Light is refracted and reflected, and by whose Vibrations Light communicates Heat to Bodies'. The same medium, being less dense within matter and more dense in empty space, is also the cause of gravity (Newton 1952: 349, Query 18; 350, Query 19).[12] But the most succinct and striking formulation of Newton's most recent escape from the problem of activity in Nature was in the concluding lines of the second edition of the *Principia* four years before:

Now it would be possible to add not a little here about a certain very subtle spirit penetrating dense bodies and hidden within them, by the force and actions of which the particles of bodies attract each other at the smallest distances, and cohere when they touch; and electrified bodies act at greater distances, both repelling and attracting small bodies near by; and light is emitted, reflected, refracted and inflected, and heats bodies; and all sensation is stimulated, and the limbs of animals are moved at will, namely by the vibrations of this spirit spreading from the external organs of sensation to the brain, and from the brain to the muscles, through the solid fibres of the nerves. But these matters cannot be explained in a few words, nor have we the sufficiency of experiments by which the laws of action of this spirit ought to be accurately determined and demonstrated. (1713: 484)

It is evident that despite the identity of words, this universal 'spirit of Nature' of the 75-year old Newton is the antithesis of the quasi-divine Spirit of Henry More. It has all the characteristics of a refined Cartesian mechanism. Must we, therefore, conclude, from these passages in the *Mathematical Principles* and the *Opticks*, that Newton has at last abandoned More and transcendental physics? Or is it the case that somehow the idea of ubiquitous, active God and ubiquitous, active spirit are to be combined? This appears to be the sounder alternative, for Newton never renounced his attachment to 'active Principles, such as are that of Gravity, and that which causes Fermentation, and the Cohesion of Bodies', the expression of which remains in the *Opticks* to the last. Moreover, Newton's most eloquent avowal of the importance of God in natural philosophy filled the pages of the General Scholium just preceding the passage about the very subtle spirit. I do not think that the spirit carries Newton one step nearer to mechanism in the Cartesian or Spinozan sense.[13] I do not see why Newton should not have imagined his new-found spirit to be the particular and universal vehicle of the divine activity in Nature. It is certainly highly improbable that he would have conceived of it as a source of natural activity independent of God, or even as a Spirit of Nature after the style of Henry More.

More's one personal appraisal of Newton in 1680 (p. 204) hints at his awareness of the different bents of their minds. We can only endorse his judgement. However deeply Newtonian scholars plunge into Newton's metaphysics, where the influence of neo-Platonism and of More especially is strongest, as well as that, maybe, of the esoteric alchemists and the religious mystics, they should not forget that Newton was the most critical of annotators, and that he used More's ideas not for More's purposes but for his own. Nor should it be forgotten that Newton was a mathematician and an experimental investigator: More was neither of these things. Newton was to consolidate the Scientific Revolution: More was not even a pioneer in that movement.

12

Conclusion

HISTORIANS' ULTIMATE sense of the role of Henry More in, or rather in relation to, the Scientific Revolution of the seventeenth century must vary directly with their understanding of the nature of that revolution. At one extreme, an historian favouring a strongly positivist interpretation, emphasizing observational and experimental discoveries and building a composite picture as a sum of particular achievements in the individual sciences, will be likely to have little or no occasion to mention More. No positive advance in any branch of science or mathematics can be attributed to him:

> It is certainly accurate and fair to say that More was interested in the new science, was a member of the Royal Society, and (meagre praise) was the most scientifically-minded of the Cambridge Platonists; it is necessary to add, nevertheless, that he had little understanding of the nature of scientific evidence, and that he conceived of science almost solely as an aid in the apologetic defence of deity. (Greene 1962: 452–3)

At the other extreme, the historian of ideas in recent years has demonstrably found More a fascinating figure, not only as metaphysician but as an author whose writings show a strong interaction with the scientific movement of the seventeenth century. That More has a proper place (if a somewhat minor one) in the history of philosophy has never been in doubt; and because no sharp dividing-line can be drawn between philosophy (in the current signification) and natural philosophy, and because the topics in which More was interested naturally take him to the border-line in question, it may well be argued that More has a place in the history of science, not least as an intelligent critic of the scientific movement of his time. A critic whose arguments had some substance in his own day, though no longer relevant to ours, must merit a place in the history of the arts, literature or science. It would not be wholly trivial, perhaps, to explore the possibilities of comparison between More and, say, Ruskin or Herbert Spencer. If the history of science is something more than a catalogue of positive achievements, if it is the ambition of certain historians to understand the scientific movement of the seventeenth century in all its aspects, to examine its failures as well as its successes, then the critics of the movement as well as its supporters must be brought under such historians' review. By 'critic' I mean here, of course, not an opponent or detractor – though the word is admittedly more often used in this pejorative sense – but an author who considers a subject or work in the light of some coherent standards of judgement; for the basis of criticism must be informed comparison.

Criticism ends and detraction begins when writers cease to measure their subject by standards of judgement but instead make it their business to press upon the reader some ideal or position of their own preference, opposed to that of the work under consideration. In the seventeenth century, for example, a great deal of the writing directed against Copernicus (and later, against Newton) was not criticism but detraction, for the authors started simply from the assumption that Copernicus (or Newton) was in error, and strove to demonstrate this point. With respect to the new metaphysics of the Scientific Revolution, however, More was a true critic; this role is plainly apparent in his letters to Descartes. Generally, More's views on time, space and spirit may be said to have been expressed in a properly critical style, though the shrill tone of dogmatic assertion begins to be heard. In his treatment of Boyle and Hooke, More ceases to be a critic and becomes a detractor, for he is concerned only to assert his own spiritualist view of the world. He has ceased to make the effort to understand the facts and ideas to which he addresses himself.

Criticism, properly speaking, is a process not a product. The decision reached as a result of the process may be adopted by posterity or it may not. Where the conclusions of a proper piece of criticism have been rejected by posterity (as with More) it will usually be found that standards of judgement, rather than the process of reasoning, have altered through time. If, tentatively, one considers More a 'good' critic of the scientific movement of the seventeenth century (one might also employ such epithets as fair, sound, judicious, well informed ...) while we ourselves are very far from accepting as true the conclusions of his process of criticism, this is certainly because the standards of his judgement, that is, his personal balance of the spiritual against the material, no longer seem appropriate to our view of things. But the claim that he is a 'good' critic could be justified only by showing that in More this balance was not inappropriate to his own time, that More was not ludicrously idiosyncratic and that (perhaps) he was really more consistent than those from whom he dissented, who seem also perhaps precursors of our own patterns of thinking.

Starting from fundamentals, it is evident that More's position as a Christian apologist, common to all the Cambridge Platonists, was not out of step with his age. Very many mathematicians and investigators of Nature were divines; some of those who were not, like Robert Boyle and Isaac Newton, professed themselves Christian apologists and natural theologians. Furthermore, More's beliefs that the universe is providential, that it was created to enable humanity to fulfil the destiny planned for it by God, and that there is (as it were) a tripartite interrelation between God, Humanity and Nature, God's twin creations, were equally universal. In some sense, then, More's insistence on the omnipresence of spirit in the universe was necessarily all but universal also, for how could a Christian providentialist *not* believe that God is continually present in His own creation – as Newton believed him to be? Again, could a seventeenth-century Christian *not* believe in angels, that is, spiritual beings active at God's behest in the physical universe? I do not think that More's talk of angels and spirits can lightly be dismissed as unusual for his own day, though the context of his discussion of such beings was already becoming unusual. Moreover, the possibility of miracles – to be denied by a Christian at his peril – surely entails not only providentiality but the activity of spirits in the universe.

As mentioned before, the crucial question in relation to all these matters is the following: do Christian philosophers believe the universe to observe consistency and regularity in its phenomena so that, as Leibniz said, any departure from this law-abiding behaviour is a miracle, or do they suppose everything to depend upon the arbitrary will of God exercised through spirits? Or to put the point differently using Newton's language, where does the natural philosophy that deals with mechanisms end and the natural philosophy that discourses of God commence? It is evident, for example, that Newton's treatment of this issue is guided by his own concerns and knowledge: he could have offered different examples of the logic of design and providence. Even after reading Leibniz's arguments against Newton in his correspondence with Samuel Clarke, one is left with the sense that for Newton the point of transition from the mechanical philosophy of Nature to the natural philosophy of God was not soon reached. Leibniz himself had to reach such a point of transition at some level. With More, on the other hand, the point of transition was very soon attained; it seems that in his view only the most rudimentary effects of Nature could be accounted for by purely materialist causation. It might be unjust to More to assert that Newton was the truer representative of his time, because More made too much of spirits. The popularity of More's writings might speak to the contrary. Features of the temper of the late seventeenth century suggest that however much mechanism commended itself to philosophers and experimentalists, it did not win the approval of the multitude.

Setting aside the individual instance of Isaac Newton, can Henry More be said to have had any definable influence upon the scientific movement during the late seventeenth century, or even upon its intellectual context? If we would establish such an influence, it must be manifest in some scepticism concerning the mechanical philosophy, though not touching *experimental* philosophy, which More always praised. His distinction is quite legitimate. The historian must look for some weakening of the authority not only of Descartes himself, but of such figures as Boyle and Hooke and of the Royal Society in general. Did More's critical writings exert such a negative influence? It is hard to discern. The satirists like Swift, the greatest detractor of the Royal Society, did not follow any example set by More. Similarly, writers upon natural religion, God and the creation, and the falsity of atheistical mechanism make little appeal to his books. Indeed, More's influence upon the development of English natural theology presents something of a puzzle. Raven has argued that John Ray's *Wisdom of God manifested in the Works of the Creation* (1691) was the *fons et origo* of this natural theological tradition, and the book was long popular in its own right; moreover, it formed the basis of Derham's Boyle Lectures in 1711–12; it supplied the background for the thought of Gilbert White and indeed for the naturalists of three generations; it was imitated, and extensively plagiarised, by Paley in his *Natural Theology*; and more than any other single book it initiated the true adventure of modern science, and is the ancestor of the *Origin of Species* or of *L'Evolution Creatrice*. (1950: 452)

Raven also traced *The Wisdom of God* back through Ray's University sermons of the 1650s to the second book of Henry More's *Antidote against Atheism* (1653), which 'supplies the sequence and underlies the contents of a large section of the *Wisdom*'. However, Raven also opined that Cudworth's *True Intellectual System* (1678) 'underlay' Ray's 'philosophy of creation' (1950: 457–60). The Preface to the first edition of *The Wisdom of God* does indeed include a general acknowledgement to More (along with Cudworth, Edward Stillingfleet, Robert Boyle and Samuel Parker) as one of the 'most learned men of our time', upon whom Ray had drawn in writing his book (1950: 452). In the text Ray also appeals often to Cicero (*De natura deorum*) and John Wilkins. More's *Antidote against Atheism* – and this work alone – is referred to as authority or precedent some nine times. Ray commended Cudworth's notion of 'plastick Nature' without mention of More's earlier contention for the active spiritual component within the universe. In his later general scientific book, *Three Physico-Theological Discourses* (1713) Ray introduced More's name twice only (pp. 331; 341): for a critical judgement of the Stoics, and the opinion that the world might come to an end through the Sun's extinction.[1]

Richard Bentley too, in *A Confutation of Atheism* (1693), amid many references to the Bible, the Fathers, classical authors (Cicero again) and (extensively) Boyle and Newton as well as other recent writers, is silent with respect to the whole group of the Cambridge Platonists. In that classic manual of both Cartesian and Newtonian anti-Cartesian doctrine, *Rohault's System of Natural Philosophy illustrated with Dr Samuel Clarke's Notes*, as translated by his brother John (1723), there is a specific but negative allusion to More; where Rohault treats of mechanics (chapter 10) Clarke notes:

Hence it is easie to explain that Paradox, which so much perplexed the famous *Dr. Henry Moor*, and other learned Men, *viz.* why a flat round Board, such as a Trencher, when it is put into Water, should rise up immediately, though the Weight of the incumbent Water be much greater, than that under it . . .

At least this proves that, many years after his death, More's polemics against the established science of his time were not quite forgotten.

It was of course unfortunate for More's fame that (as with Clarke's book) late British anti-Cartesianism was overwhelmingly Newtonian. It might be said of More that when he attacked the mechanical philosophy as an epistemology of Nature he was disregarded; and when he attacked specific-ally Descartes's philosophy of Nature his candle was lost in Newton's greater light.

For example, all the writers examined above (as well as many others) rebutted Descartes's concept of the animal as a machine; not one noted More's early and sustained rejection of this branch of mechanism. The Cartesians were at least logical in drawing their conclusions from man's unique possession of a rational soul. Why should the movements of an animal be less automatic than those of a clock? Why should not a plant grow after the same manner as a crystal? More was equally logical in asserting the antithetic proposition that spirit (not soul) is everywhere: in animals as in human beings, in inanimate Nature as in plants. He rejected the illogicality of the possibly tactful compromise that held inorganic Nature to be mechanical, organic Nature vital. That animals should be machines seemed absurd to More:

> To be sure, how can it happen that the voices of the parrot or the magpie imitate our own, unless they hear our speech and perceive it [with organs of] sensation? But, you will say, they do not understand what those sounds mean to a human, which they babble out by way of imitation. However, do they not understand what they want, that is, the food which they win from their master by this trick? Therefore they think that they are begging food for themselves, because they gain their object each time by their talkative-ness. And to what end, I ask, is that attention and listening in singing birds, which they display, if there is neither sensation nor consciousness in them? And for what use is the astuteness and good sense of the fox and the dog? Why do threatening words and menaces restrain wild beasts? When the household dog steals something, why does he slink away secretively as though with a guilty conscience, congratulating himself on meeting no one as he cautiously and hesitantly slips away to a distance, with muzzle down

and averted, prudently mistrustful lest in redress of the crime a penalty be exacted? How can this be, without inner consciousness? (Descartes 1974: 244, my translation)

More had cogent philosophical as well as experiential reasons for rejecting the Cartesian mechanistic account of living things, and exactly analogous reasons for rejecting mechanism in the inorganic creation. In his own strange way, More was consistent: he denied that anything *active* could be truly mechanical, that is, wholly deprived of the action of spirit. He is consistent too in affirming that there is a unique source of activity in all Nature, in the inorganic realm as well as in the organic, that is spirit, whose function it is to energize both the flight and the fall of the sparrow. To the mechanists who nevertheless preferred vitalism against Descartes such consistency was denied. The rejection of spirit in the inorganic world forced them to postulate 'living force' in the organic world, which is in effect spirit under another name. The infusion of life into inert matter, the nature of the body–soul interaction, the drawing of a boundary between life and non-life – all such problems presenting themselves to non-Cartesian mechanists did not exist for More. He had severed the Gordian knot of dualism with his spiritual sword.

If the willingness of the strict Cartesians to extend the tenets of mechanism to the organic creation (but to the exclusion of Man) avoided some of the difficulties associated with vitalism, they did not escape the problems of the 'as if'. By this I mean that for whatever reason – whether he believed that natural philosophy could be at best a probable story, or because he feared the *odium theologicum* or merely for reasons of convenience – Descartes did not pretend to describe the real origins and structure of the world that God had made, but to write of a universe whose properties are identical with those of our own. Hence, as Koyré puts it, 'Descartes's God is perhaps not the Christian God, but a philosophical one' (Koyré 1957: 122). Descartes did not philosophize the Book of Genesis, as More was to do. Descartes claimed only to describe and analyze a world that corresponded in every detail of manifestation with the world created by God, not that actual world. Descartes's *Monde* is the world *as if* we could determine what God had really done and made, which (of course) we cannot do. Now if these statements of Descartes are to be regarded as no more than a specious shield against criticism, then their serious value is slight and they need not be taken as representing his true beliefs. On the other hand, they may be read as expressions of a genuine scepticism about the philosophic possibility of knowing reality, or as an assertion that (after all) if the world of 'as if', the model, corresponds to our experience in every particular, it is needless to seek reality. However, it seems to me that by contrast the philosophers of the English school from Bacon to Newton scorned any such prevarication; they meant to investigate the world that God had really created (hence the importance to them of the Biblical guidebook to Nature) and to analyze the world as it really is. Their position was shared by the least theological natural philosopher of this age, Christiaan Huygens, who for all his neo-Cartesian mechanist proclivities rejected the physics of 'as if' as *'un beau roman de physique'*. Leibniz, on the other hand, adopted without scruple the Cartesian 'as if' from his *Hypothesis physica nova* of 1671 onwards. Leibniz admits that God and spirit exist in the real universe, that miracles have occurred and so forth but all these are barred from his philosopher's world of 'as if'. As with Descartes, the rational world of Leibniz, excluding the Christian's knowledge of what God has done as readily as the naturalist's knowledge (and ignorance!) of what exists in the real universe, is necessarily a world of 'as if'. For only the world of 'as if' can promise the possibility of *complete* knowledge to the philosopher. If his knowledge of the world is less than complete (in principle, at least) then it is to that extent also less than fully rational, since incompleteness of knowledge implies a defect of reason. But the empirical philosopher, like the natural theologian, confesses that his knowledge of the universe is, and must ever remain, less than complete because part of the universe is hidden from him in distant space and part in remote time. In this way there is a profound link (in the seventeenth century) between empiricism and religion; neither can claim the logical completeness of rationalism.

Now Henry More was not a man who could ignore the evidence concerning God's intentions and deeds which he believed to have been furnished to humanity by God himself, nor was he (in the long run) interested in a purely rational world. It may seem odd to say of a man who professed himself a Platonist that he was ravished by the wonder of scientific discovery, yet this is the inescapable conclusion to be drawn from More's philosophical poems and later works. The mysteries of God's creation, which human reason unaided was powerless to decipher, might be glimpsed piecemeal by the exploring eye. The patient experimenter might stumble towards the true vision of things where the acute reasoner, like Descartes, deludes himself by following the will-o'-the-wisp of his own mind. More never rejoiced in Descartes as the reviver of Platonic reason, but as reformer of the ancient atomist philosophy, 'that sublime and subtil Mechanick'; Platonic reason and Cartesian rationalism were always distinct in his mind, and for the latter he felt much distrust. Where Descartes can only operate in the world of 'as if' (which is to More the world of the shadows seen in the cave), More's 'platonick' philosophy of the world is for him *real*, neither a romance nor a working model. He joins hands with the natural theologian on the one side (for whom the world is real because of the evidence of God's creating it) and the experimentalist on the other (for whom the world is made real by investigation).

The Christian rationalist philosopher must admit, even proclaim, the real existence of the supernatural yet exclude it from the world of 'as if' which is his philosophical concern. Neither Descartes nor Leibniz nor Malebranche can hesitate to affirm the truth of miracles, but because miracles are not features of the customary order of Nature they fall outside of their province. The experimentalist can set aside miracles for an obvious cognate reason, but (in the seventeenth century at any rate) he had greater flexibility than the rationalist in that he can, indeed must, on occasion admit ignorance of causation. To the rationalist (as Leibniz wrote to Clarke) the admission 'I do not know the cause' is equivalent to 'I see a miracle.' To the mind of the experimentalist the so-called occult cause is not appalling: the truth of things can appear to them (in Newton's words) 'by Phaenomena, though their Causes be not yet discover'd'. To declare that an effect or indeed a 'general Law of Nature, by which the things themselves are form'd' is manifest in Nature though its cause is unknown is not to declare that the cause is unknowable or (so to speak) lies outside the ordinary course of Nature. As Newton continued, echoing More's words already quoted (p. 182):

> To tell us that every Species of Things is endow'd with an occult specifick Quality by which it acts and produces manifest Effects, is to tell us nothing: But to derive two or three general Principles of Motion from Phaenomena, and afterwards to tell us how the Properties and Actions of all corporeal Things follow from those manifest Principles, would be a very great step in

Philosophy, though the Causes of those Principles were not yet discover'd:
... (Newton 1952: 401–2)

For all Newton's cautious confidence, the possibility cannot be excluded *a priori* that the cause will not become known by further investigation or rather that no physical or mechanistic cause will become known; and that because of the dependence of the natural world upon the supernatural, which Newton elsewhere admitted but did not wish to allege on this occasion, the cause of a natural event may exist in the supernatural world, no longer excluded by the 'as if' stipulation.

A case could be made that More was pursuing this line of thought, adumbrated long after his death by Newton – one that seems eminently proper to the Christian seventeenth century – to its logical limit. If the spiritual or supernatural world exists at all, if the physical universe was created to serve the purposes of higher, spiritual beings, is it not reasonable to hold that there must be an interaction between the realm of spirits and the universe of matter and that this interaction is so regular as to constitute some degree of dependence of the latter upon the former? The only alternative position which is equally logical, commending itself to no one in More's time, is to deny the supernatural altogether.[2] It may be said that More pushed his supernaturalist philosophy to absurd lengths and was (as everyone has said) absurdly credulous. But the trend of his thought was not out of keeping with the general presuppositions of his age.

It was, on the other hand, wholly counter to the mechanical philosophy, which has rightly been reckoned the central intellectual pillar of the Scientific Revolution. The object of the mechanical philosophy was precisely to exclude from accounts of Nature all terms that were not strictly definable and complete in themselves; such terms as time, space, mass, shape and motion only were to figure in scientific discussion at the fundamental level. In principle the basic concepts of scientific thought were to be susceptible to mathematical development; this Cartesian ambition, only slightly realized by Descartes himself, was achieved by Newton. We must note, however, how far Newton's success in mathematizing Nature was from denoting him as an intellectual heir of Descartes. For his mathematical success followed only from his abandoning that mechanical philosophy (in the original sense) which had so seduced Descartes into the world of 'as if', the swamp of hypothetical, qualitative mechanical models with which he had attempted to replace the reality of Nature.[3] 'In most of the Writings of the Philosophers,' according to John Keill,

> there is scarce anything to be found besides the name. Instead whereof, the Philosophers substitute the Figures, Ways, Pores, and Interstices of Corpuscles, which they never saw; the intestine motion of Particles, the Colluctations and Conflicts of Acids and Alkalies, and the Events that thence arise, they relate so exactly, that there is nothing but a Belief wanting in the History of Nature, as often as they set forth the Miracles of their subtile Matter . . . (1726: Preface)

The early Newtonians were well aware that their 'true philosophy' (as Keill boldly described it) had thrust aside this hypothetical mechanical philosophy, which had in fact (as Newton proved) run counter to the genuine mathematical science of mechanics. Errors came about because Descartes, though a 'famous Geometer', 'that he might accommodate him to the idle and common Herd of Philosophers, made no use of Geometry in his Philosophy' with the consequence that 'although he pretended to explain all things mechanically by Matter and Motion, yet he introduced a Philosophy, which was as contrary to the true Laws of Mechanicks, as was possible' (1726: Preface). Keill traced the line of true philosophy from Archimedes through Roger Bacon to Galileo and the scientific societies, among whose chief ornaments (in addition to the incomparable Newton, who had accomplished ten times as much as all his predecessors together) he reckoned Huygens, Boyle, Wallis, Halley and David Gregory, whose exposition of the Newtonian 'Mechanical Principles' to the motions and inequalities of the planets 'will last as long as the Sun and Moon endure'. Gregory, a fellow Scot, was Keill's teacher and first patron.

Though Keill is insistent upon the methodological change in the mechanical philosophy effected by the Newtonian school, he is less clear about the other shift in its character. He sees the crucial significance of the mathematization of the mechanical philosophy without understanding how this was made possible by the change from the kinematic to the dynamic approach. The reason for this lack of clarity about the role of forces in Newtonian physics is, of course, that reluctance to exclude the notion of aetherial mechanism which is characteristic of Newton's own cautious expressions. The Newtonian philosopher claimed the right to use *force* as an unknown quantity – in Keill's expression, like the x and y of algebra. Its reality was made known by its effects although its nature and cause were unknown. Reluctant to admit that force is an 'active principle' in matter or space, Newton alleged – to protect himself – that all forces might after all prove to be mechanical in the Cartesian sense, though not after the Cartesian fashion.

In this way, then, Newton and his followers were less open and honest than Henry More, who proclaimed his root purpose of denying in principle that forces in Nature could be mechanical and so, at bottom, without cause. There was no ambiguity in his position.

It is clear that Newton's refusal to align himself with More in a position that he regarded as unbefitting a natural philosopher was quite deliberate, even though his general sympathy for More's idealism was so strong. That the reading of More, like that of Walter Charleton, was a persuasive factor in Newton's adoption of atomism as a young man no one doubts. Westfall, after remarking that More was an influence throughout the early *Questiones Philosophiae*, continues with the further point that 'the influence of Henry More was particularly strong in those passages intended to refute the possibility of a material order autonomous and independent of spiritual control' (1971: 326; 337). And he plausibly conjectures that precisely this lesson learnt from More was an effective reason for the devout young Newton's violent rejection of the materialist philosophy of Descartes. But, while departing from Cartesianism, Newton equally rejected More's own spiritualist philosophy. In a sense, for More's spirits Newton substituted active powers, principles and forces which, though not essential to matter existentially (for it is possible to conceive of matter lacking such forces) nevertheless do belong to the mechanical order of Nature. While it is true that Newton might find the only possible cause of force in God's will, it has to be added that he was equally willing to imagine that matter too exists because of God's constant will, and so there was really no distinction in this respect between his concepts of them. Matter, force and consequently motion are as much divine as mechanical, and vice versa. I do not know whether Newton at any stage considered himself to be reconciling the spiritualism of More with the mechanism of Descartes, but this seems to be in effect what he did. 'Force' is a new category of ontology, one that could not be claimed for his own camp either by the one or the other.

Perhaps nothing marks more sharply the passage from the materialist world of Descartes and Leibniz, a world full of matter, to the dynamic world of Newton than Newton's insistence on the minuteness of the quantity of solid matter in the universe; this was a point of which Voltaire was fully conscious. In *Opticks* Newton argued that if the first order (spherical) particles are in contact, closely packed, within a given volume of space, so that the space occupied by solid matter is one-half of the whole, and second-order particles are similarly arranged within those of the first order, then the ratio of vacuum to solid matter is $(2^n - 1)$, where n is the number of orders of the particles (Newton 1952: 268–9).[4] If, in accordance with the general theory of attractive force, the particles are not required to touch each other so that the initial proportion of solid to vacuum is less (say $1/m$) then the total amount of solid becomes rapidly much less as the number of the orders of the particles increases $(m^n - 1)$. Keill is one Newtonian who makes much of this immateriality of matter; it is hard not to believe that Henry More would have been delighted by such computations. For it is an easy inference that if the universe is only thinly populated by matter – a point argued at some length by Newton in the *Queries* – the greater must be the importance of that in space (whatever it may be) which is not matter.

In counterpoint to the long-lasting influence upon Newton of More's metaphysical philosophy was the no less decisive impact of Cartesian scientific philosophy. This matter too has been fully discussed by historians and may here virtually be taken for granted: as for many or most Englishmen of the 1660s, Cartesianism was the 'paradigm' of science, 'the philosophy of the common sort' in Newton's phrase. Like More earlier, Newton read, learned, was deeply impressed, but after years of reflection put Descartes behind him to go his own way. Not until the late 1670s, however, was Newton ready

> to reject the fundamental tenet of Descartes's mechanical philosophy of nature, that one body can act on another only by direct contact. Newton's mathematical papers suggest that only a wholesale repudiation of his Cartesian heritage would enable him to take that step. The repudiation determined not only the content but also the form of the *Principia*. (Westfall 1980: 381)

Henry More, we may suppose, had some role in bringing Newton to read

Descartes in the first place, and had a more obvious influence in assisting Newton to repudiate Cartesian mechanism (to repeat Westfall's term). In fact, if we have little enough evidence to bear out the usual contention that More had some part in interesting his fellow countrymen generally in Descartes's writings, we have virtually none so far as Newton as an individual is concerned. If one had to make a guess, it might seem more likely that it was Barrow who induced Newton to 'look into' the Van Schooten enlarged Latin edition of Descartes's *Géometrie*, as also the *Dioptrique* and the *Météores*. The significance of More's writings, so far as Newton is concerned, is confined to the fields of theology, metaphysics and cabalism. We may well imagine that Newton found in More confirmation of that tendency to view matters symbolically that was surely entrenched in his own mind. In the open expression of some of his most recondite (if not profoundly original) thoughts Newton turned again and again to the idea that truth (one and simple) is concealed beneath appearances (manifold and deceptive); so he wrote in the concluding lines of the last edition of the *Principia* of that universal and elastic spirit that (he suggests) is the chief agent of coherence and gravity, optics and electricity, as well as of nervous physiology – this certain most subtle spirit 'pervades and *lies hid* in all gross bodies' (Cajori 1946: 547, my italics). The same spirit appears in the *Queries* concluding *Opticks*, where even more than in the *Principia* Newton treated the First Cause as a cabala, secret knowledge of it coming slowly to humanity from the true philosophy: 'does it not *appear from Phaenomena* that there is a Being incorporeal, living, intelligent, omnipresent, who in infinite Space, as it were in his Sensory, sees the things themselves intimately, and thoroughly perceives them, and comprehends them wholly by their immediate presence to himself' (Newton 1952: 370, my italics).[5] The philosophical astronomer sees through the variety of bodies and multiplicity of motions in the universe to discover the 'wonderful Uniformity in the Planetary System [that] must be allowed the Effect of Choice' on the part of the supreme Being, just as the student of living creatures in all the range of their forms and habits discovers

> the Uniformity in the Bodies of Animals, they having generally a right and a left side shaped alike, and on either side of their Bodies two Legs behind, and either two Arms, or two Legs, or two Wings before upon their Shoulders, and between their Shoulders a Neck running down into a Back-bone, and a

Head upon it; and in the Head two Ears, two Eyes, a Nose, a Mouth, and a Tongue, alike situated. (1952: 402–3)

Since Nature is neither a chaos nor an order that came about by chance, but the 'effect of nothing else than the Wisdom and Skill of a powerful ever-living Agent', the business of the investigator of Nature is to decipher from the message written in numbers in the Book of Nature the traces of the First Cause (this is analysis) and then to reconstruct his plan and purpose (this is synthesis). It is well known that in his studies of ancient history, of classical mythology, of religion and of alchemy, Newton always sought beneath the superficial meaning of texts for the keys which would enable him to decode the hidden true meaning. This was all in the tradition of Henry More and Joseph Mede.

The breaker of the codes in Nature never has more need of all his skill and ingenuity than in penetrating the mysteries of force, for this is hard to discern beneath the phenomena of motion and change. More than a century passed after the publication of the *Principia* before any force other than gravity submitted itself to mathematical analysis. It was a bold step for Newton to guess that the dynamism of Nature works in the same way in the (then) qualitative observable effects of optics or chemistry, as in the directly ascertainable motions of planets and tides, though (it should be added) Descartes and Boyle had prepared the ground for Newton's step.

> I therefore propose [wrote Newton in a draft Preface for the first edition of the *Principia*] the inquiry whether or not there be many forces of this kind, never yet perceived, by which the particles of bodies agitate one another and coalesce into various structures, For if Nature be simple and pretty conformable to herself, causes will operate in the same kind of way in all phenomena, so that the motions of smaller bodies depend upon certain smaller forces just as the motions of larger bodies are ruled by the greater force of gravity. It remains therefore that we inquire by means of fitting experiments whether there are forces of this kind in Nature, then what are their properties, quantities and effects. For if all natural motions of great or small bodies can be explained through such forces, nothing more will remain than to inquire the causes of gravity, magnetic attraction and the other forces. (Hall and Hall 1962: 307)

We should note that Newton's expectations of the analogy between the microscopic and the macroscopic worlds are different from those of Descartes and More in an important respect. Both of these philosophers thought that their source of activity in the universe (transfer of motion, or Spirit of Nature) would always function in the same way. On the other hand, Newton finds it probable that the microscopic or interparticulate forces will be distinct in their properties from the forces operating between large bodies, such as gravity and magnetism; he envisages the possibility that electricity as a short-range force may have different properties from electricity as a long-range force. 'Force' is not like 'motion' or 'spirit' an unqualifiable universal explicans: experimental research must be undertaken to discover its variety of forms and categories. To revert to Toulmin's terminology (1959), we might say that Newton here introduces a 'useful' concept as distinct from More's 'vacuous' notion of spirit. 'Force' is not and never was a metaphysical concept uniquely; it is also a mathematical and an experimental concept. It would not have been possible for More to explore mathematically and experimentally the application of his notion of the Spirit of Nature, even had he thought it necessary so to do.

When More tried to enter the experimental field he made a fool of himself, as Boyle told him. What we must remark of Newton is that just as he found it intellectually essential to state certain metaphysical foundations of his physics so, equally, he could advance from these metaphysical postulates to the level at which experiment and calculation become possible.

Nevertheless it has been suggested that here too More may be found to have exerted an influence, at any rate upon Newton, for More was somewhat less than a perfect rationalist. He understood in principle as well as Boyle did (against the rationalist Spinoza) that the claims of physical truth must be compared with those of logical truth (Hall and Hall 1964). While all historians agree that the most important of More's criticisms of Descartes are at the level of metaphysics, they reveal 'in More an acute sensitivity to the metaphysical problems that Cartesian physics left unanswered' (Mamiani 1979: 61, my translation). More chooses to differ from Descartes not only or even chiefly on metaphysical generalities but on points that concern the physical nature of the universe; its infinity, its composition of atoms and void, the motion of its parts. Mamiani insists on the importance of More's distinction, in opposition to Descartes, of logical necessity from physical necessity in the discussion of such questions. The course of human thought can impose no constraint upon the nature of real things, or their causal relationships:

> All these points strike against the mechanistic reduction effected by Descartes in his physics and will provide the basis for new attempts to investigate Nature which, by not complying completely with Cartesian mechanism, will modify it by the significant introduction of non-mechanical concepts, such as attraction and the electric spirit. (1979: 88–9)

Thus Mamiani (like many others) looks directly forward from More to Newton. He also suggests that More's criticism of Descartes reflects already something of the English empiricist spirit, 'a new cultural sensibility in the Cambridge context, and a little later that of the Royal Society.' 'Thus the truly fundamental point of divergence between More and Descartes is the clear separation effected by the Englishman between logical necessity and physical necessity, which for More renders indispensible recourse to experience in order to understand the nature of matter' (1979: 74; 82). For, if experience or knowledge of reality were irrelevant, it would be pointless for More to object against Descartes that extension is not a characteristic of matter only, and that extension does not by definition comprehend the ideas of impenetrability and tangibility, properties whose association with matter we learn only from experience. 'By attacking Descartes on this point, More makes a radical defence of the legitimacy of the positive use of experience in the formation of scientific concepts' (1979: 74).

These thoughts may perhaps seem to assimilate More's epistemology too closely to that of a group with which More did not comfortably belong: More was never to accept the kind of natural philosophy that was favoured by the Royal Society. But it is relevant that he did not object to the evidence of pneumatic or hydrostatical experiments as such; his quarrel was with the mechanistic interpretation of them. More's contention – one that recent writings on the philosophy of science have made more commonplace to us than it was in More's day – is that experiments cannot be made to pronounce unambiguously and positively in favour of some set of ideas, for the lessons drawn from experiments are inevitably conditioned by preconceived notions and presuppositions. There can be no atheoretical interpretation of an experiment. More makes, early in life, the same point against Descartes that late in life he makes against Boyle and Hooke: when we make an assertion about the structure of the natural world (always a matter of inference, never something we can simply 'see' or handle) we must be scrupulous to distinguish those elements in our assertion which derive from experience (the hardness of matter, the weight of fluids) from those elements that derive from our own intellects (the notion of extension, the notion of elasticity). The practice of experimentation is (More understands) above all important in clarifying this distinction and so in permitting us to examine our thoughts more self-consciously and exactly. This is a function which he takes to be essentially negative, enabling us to falsify those ideas of Nature which are plausible but incorrect (such as those of mechanism) and again by exclusion of these confirming those ideas that are valid, such as the notion of spirit. For such reasons More welcomed the hydrostatical and pneumatic paradoxes published by Boyle. Similarly, when More seems, against Descartes, to argue for experience against intellect − physical necessity against logical necessity − it is for the negative purpose of falsifying Cartesian mechanism; More is certainly not intending to deny that there is another world of thought − the Platonic − on which we may confidently rely. It would surely be a mistake to regard More's use of the argument from physical necessity as weakening his commitment to Platonic idealism.

It might in any case be argued that in so far as More taught the value of experience as against reason, he was in some danger of preaching to the already converted. For the height of More's literary success and intellectual influence was during the two decades 1655–75, roughly speaking. His publications and teaching before 1655 may have carried weight, especially the *Philosophical Poems,* but this has yet to be proved. Equally, in the history of philosophy the productions of More's last years count for little. Moreover, as is evident from what has gone before, from 1678 onwards the massive systematic scholarship of Cudworth tended to cast a shadow of oblivion over More's books, precisely in those areas where they might have been most effective, that is, in the defence of Christian atomism, of the truths of natural religion, and in the Hylarchic Principle or 'plastic nature'. By 1655, when More was past 40, the founders of the Royal Society who were his contemporaries (*grosso modo*) were already formed intellectually. Their minds had been shaped by other and older influences − those of Paracelsus, Bacon, Galileo, Kepler, Harvey, Descartes, Gassendi . . . However we may define and interweave the various groups that are associated with Gresham College, Oxford University or Samuel Hartlib, it is at least clear that chronology itself excludes the possibility of More's influence upon the impressionable, formative years of their members. And perhaps he was in the wrong place too, at least during the Commonwealth years. It would be pointless here to enter into the debate about the cultivation of mathematics and science in the English Universities, since partisans on both sides agree that, in the middle years of the seventeenth century and more particularly after 1649, Oxford enjoyed a predominance of wit and vigour, attracting men like John Wallis and Seth Ward from Cambridge.[6] At no time in his life did More desire to draw into his orbit a group of students and investigators, as Wilkins did at Oxford. Years later a correspondent, Giles Alleyn, was to write to More:

> I wish yt some p[er]son of worth or other that hath where withall would erect a lecture for experimentall philosophy [at Cambridge] and would give such a yearly income towards it, as yt all things needfull for ye trying of experiments might be had: wch if it were done, I should not doubt but yt ye University would quickly farre surpasse ye royall society. (25/1/1674, Christ's College MSS)

But such things were quite outside the pattern of More's life. In so far as natural philosophy became in the England of the 1640s and 1650s co-operative, experimental and mathematical, and extra-academic, More had no part in it.

Therefore More's chief impact could only have been upon men almost a generation younger than himself – such men as Nehemiah Grew (b.1641), Isaac Newton (b.1642), John Flamsteed (b.1646) or even Edmond Halley (b.1656). But these men who constituted the second generation of the Royal Society were also subject to the powerful teaching and example of its founders and of continental mathematicians and philosophers, whose work was by no means wholly cast in the Cartesian mould. The various branches of science, rapidly evolving, were acquiring a momentum and pro-fessionalism of their own, weakening the influence of general philosophers like More.

Perhaps, after the publication of *The True Intellectual System*, Cudworth's influence was the greater because he was a more positive thinker than More. For, as it seems to me, in the latter decades of More's life, his critical function came increasingly to lie in decrying all facets of the mechanical philosophy and its atheistic tendencies, rather than in elaborating his own spiritual alternative, developed only in his earlier books. Many readers of More must have been puzzled by his incomprehension of, and hostility towards, the important gains in the understanding of Nature that contemporary natural philosophy was achieving. One correspondent of More's pointed out to him that

> although neither Cartesius, Hobs nor Galileo have found out the true Mechanicall solutions of these Phenomena of Nature; yet it doth not follow that there are none any more than it doth: That there was no such place as America before Columbus discovered it. I shall not deny that these men had & have as much wit, to invent arguments; and reason to infer what might follow from their inventions; and therefore were & are as likely to salve the Phaenomena, as any other; but no man is omniscient: and another that is inferiour to them, being helped by their labours, may discover something that they have not ... As a dwarf upon a gyants shoulder may see farther than the gyant himself. (H. Hyrne to H. More, 19/8/1671, Christ's College MSS)

The writer of this letter was Henry Hyrne, of Parson's Green near Fulham, who took a great interest in the explanation of the tides. He was a close friend and frequent correspondent of Henry More, who had recently presented him with a copy of *Enchiridion Metaphysicum* (see Hall and Hall 1969: 592–3 and Gabbey in Hutton 1990: 26). Provided that the philosopher will acknowledge the true existence of incorporeal beings, Hyrne asks, where is the fault in the mechanical philosophy? For,

> though one, or all of those Phaenomena [which More had claimed were not to be explained mechanically] may now or hereafter be solidly demonstrated from their [mechanical] principles; it will no more make against the existence of Incorporeall beings, than a demonstration of the succession of the day & night, upon the supposition that god causeth the earth to turn about its own center.

Hyrne declares himself to be of the same general opinion as More – that Descartes's mechanistic philosophy is far less compelling than its author supposed, and that an incorporeal agent in Nature must be admitted – but his letters show considerable reluctance to follow More down the treacherous path that would lead to the exclusion of *all* mechanical explanations from natural philosophy. Hyrne saw a genuine weakness in More's critical position: on what principle was a line to be drawn between allowable and forbidden mechanical hypotheses? More's own writings leave this question very vague.

If, as many recent writers would claim, Newton was More's pupil in evading Cartesian mechanism by his concept of force, then it is perhaps fair that Newton too was caught by the same difficulty, which in his case may be expressed as that of drawing a line between natural philosophy and theological philosophy. Where, at what point, is one man's *Ignoro* remediable, at what point in principle irremediable? For only beyond that point is it necessary to cry *Ignorabo* and invoke the intervention of spirits or of God. If Newton had really asserted that the nature, origin and manner of action of forces are in principle unfathomable by humanity – but it is by no means certain that he meant to declare this as his firm opinion – then this would have been tantamount to a declaration that no mechanical hypothesis of force-action can be framed (a view rejected by such Newtonians as Fatio de Duillier and Bryan Robinson). In that phrase a line would have been clearly drawn, but Newton hesitated to draw it, presumably (in part) swayed by just such arguments as Hyrne brought against More. Another investigator might see further.

Thus our discussion returns once more to the crucial case of Newton. No historian has as yet, I think, made out a case for More's exerting any comparable degree of influence over any other figure even of far less significance in the history of science. Nor is he known to have dissuaded anyone from pursuing science after the examples of Boyle, Hooke or Glisson. The general opinion among historians of the seventeenth century is that More's blatant spiritualism and his credulity with respect to ghosts and witches tainted his later metaphysical writings and put him outside the scientific pale.

For the Platonists, especially Henry More, science was important largely to the extent that it confirmed the existence of supernatural powers. ... It is significant that More turned away from Cartesianism not because he was dissatisfied with Descartes's scientific findings but because of their implications for metaphysics and religion. More's interest in witchcraft can be seen as part of the same phenomenon – the attempt to find in nature proof of the supernatural. (Shapiro 1969: 144–5)

Statements of the same purport, which contain a good measure of truth though one would wish to amend them in detail, were not uncommon around 1700; Leibniz's opinion was not dissimilar. Criticism did not greatly disturb More. When Johann Christoph Sturm of Altdorf, founder of the *Collegium Curiosum sive Experimentale* (1672), attacked the Spirit of Nature in an *Epistola ad V.C. Henricum Morum de Spiritu ipsius Hylarchio* (1676) More wrote quietly to Lady Conway, 'He is a good ingenious man and well seen in his way. But ... the Sandalphon will pricke up his eares for all this and that truth is invincible' (Nicolson 1930: 425, 26/2/1676).[7] Likewise of framing his reply to Sir Mathew Hale (More's *Remarks upon two late ingenious Discourses*, 1676) 'in answer to what concernes me in his two late bookes of Hydrostaticks and the Torricellian experiment', More wrote to Lady Conway, 'I hope I shall convince him of my principium Hylarchium or spiritus Naturae as well as I have Henricus Regius' (1930: 389). More was as confident of the reality of the Spirit of Nature as of the existence of ghosts and apparitions, of which (he declared) it was unnecessary for him to probe the truth of the stories that circulated (1930: 270, 23/3/1666). But this very confidence alienated everyone with a shred of the critical faculty in his being.

One cannot but feel that Newton publicly dissociated himself from More for just this reason. The relationship between the thought of Isaac Newton and that of Henry More is very much a twentieth-century discovery, diffused by A. E. Burtt and Emile Meyerson, and since their day strengthened by the study of Newton's private papers.[8] At the beginning of this century the notion that Newton was a metaphysician as well as a mathematical and experimental philosopher appeared strange, though now it seems equally hard to understand how one seeking to reconstruct the system of the world could avoid discussing space and time, topics falling into the realm of metaphysics. Newton could not be content with a merely empirical or mathematical account of Nature. It is now a commonplace to affirm that at the philosophical level Newton's

ways of thinking were in part the outcome of the neo-Platonic doctrines he had learned as an undergraduate from Henry More. The Cambridge Platonists found it necessary to allow for the regular operation of God in the world and for the endless variety of activities in living things. To the passive principle of matter they added all sorts of active principles; without them, they argued, animation and purpose could not appear. This argument was to be echoed by Newton over the years ... (McMullin 1978: 43)

This statement must be interpreted as tracing to the Cambridge Platonists the root of Newton's crucial departure from the natural philosophy of Descartes; the transition from kinematics to dynamics.

It is true that the writer (Ernan McMullin) goes on to suggest that alchemy gave Newton his most intimate acquaintance with 'the turbulence of natural process', and increased his 'conviction of the presence in Nature of perpetually working active principles', but this hardly reduces the importance of Platonism (in his view) as the chief source of Newton's *conceptual* scheme in which matter is inert and non-matter is active.

Wherein, then, do the active principles reside, what is non-matter in Newton's view? Here difficulties of terminology become acute. Not only does the word *spirit* bear many meanings but *aether* too may be regarded as a special state of matter (as by Descartes and some nineteenth-century physicists) or as non-matter, when it may be more or less equivalent to spirit. One extreme view, expressed by I. B. Cohen in 1958, but since (it seems) abandoned by him, is that the Newtonian aether:

> a central pillar of his system of nature ... was composed of particles that mutually repelled one another or that were endowed with a centrifugal force. The aether was imponderable, odourless, tasteless, and colorless, but had certain implied properties of rigidity so as to support undulations, such as those which were a concomitant part of optical phenomena and also those allied with the transmission of sensations to the brain ... the aetherial medium was capable of causing gross bodies to move toward one another according to the law of universal gravitation. (p. 7)

However, the historian adopting such a solution to the Newtonian problem has many difficulties to surmount. The documents indicate that Newton speculated on different occasions about two or more types of aether, which the hypothesis of a single aether fails to distinguish. Secondly, it is hard to see how an aether which approximates to the 'subtle fluids' of eighteenth-century natural philosophy, possessing the properties of elasticity, rigidity and particulate structure, and in addition (perhaps) that of mutual repulsion or rotation, could be said to be in any strict sense immaterial. Nor can any such form of aether hypothesis escape the problem of infinite regress; if the aether-particles give force to material particles, whence in turn do they derive their own force? Such an aether seems to leave us with all the problems of the origin of the properties possessed by matter which it is supposed to resolve.

On the other hand, the view of the older French authors (and others) that the — or rather, a — Newtonian aether should be identified with Henry More's Spirit of Nature seems equally questionable. Emile Meyerson wrote in 1936:

> when one reads Newton's words about the 'very subtle spirit which is the cause of the phenomena of cohesion between the particles of bodies, of electricity and light, as well as of sensation and animal motion', he should be careful not to take this passage too seriously. It is not Newton's voice speaking, but that of Henry More, a philosopher forgotten today, perhaps, but who was for all that well known at the time. (Metzger 1938: 75, n.6)[9]

Now no great objection need be raised against the somewhat nebulous idea that Newton's notion, expressed in the last lines of the *Principia* (1713), of an 'electric and elastic spirit'[10] was inspired by More's notion of a Spirit of Nature, though a little close examination will show that these two 'spirits' are really very different things. One of the most important functions of the Spirit of Nature was to be the cause and agent of gravitation. But Newton carefully distinguished *his* spirit from the question of the causation of gravity, which he had discussed in the previous paragraph. This point is often overlooked.

Hélène Metzger, taking Meyerson in the strong sense, went on to be more specific about the near-identity of the two spirits: 'Newton was very susceptible to the charms of certain mystical doctrines; he was receptive of suggestions springing from very diverse mental attitudes, and he particularly appreciated the active "Spirit of Nature", the means through which Henry More believed God to accomplish his works' (1938: 57). She then pointed out, correctly, that Newton was to express a similar concept to that of the closing paragraph of the *Principia* in *Opticks*, Queries 18 to 22, extending his statements to give a much fuller account of how the medium (as it is now termed) brings about a variety of optical phenomena and, in addition now, those of gravitation. Thus the change from 'spirit' in 1713 to 'aetherial medium' or plain 'aether' in 1717–18 is accompanied by not only a detailed working out of the hypothesis but its extension to celestial mechanics.

Hélène Metzger then observed, justly, that Newton vacillated between two types of thought in natural philosophy which are discordant and possibly incompatible: 'for attraction is either an unexplained, first-level phenomenon resulting immediately from the will of God, imposing itself upon the material world, or else it might be that this will of God is propagated mechanically by means of a universal spirit' (1938: 58).

We have now reached a point of confusion. Let us pass over the textual inaccuracy involved in the assertion that Newton 'appreciated' More's Spirit of Nature (we have no evidence of this, and some against). Metzger was surely mistaken to link mechanism with universal spirit, and (by implication) to associate More's influence with the latter, rather than with the will of God acting directly or indirectly. To my mind, the various antithetic 'Newtonian' hypotheses should be expressed as follows:

1 gravitation is caused by the will of God, acting either (a) directly upon matter, or (b) indirectly upon matter through the Spirit of Nature; OR

2 gravitation is caused by a material aether which is either (a) a dense, close-packed medium, or (b) a medium exceedingly rare and elastic, offering little resistance to the motions of heavy bodies.

My reading of the evidence, contrary to what has sometimes been suggested, is that so far as Hypothesis 1 is concerned, Newton's preference in relation to gravitation was for branch (a), while More's philosophy was committed to branch (b). Newton's unique allusion to More in print (though under the guise of anonymity) discounted the Hylarchic Principle (p. 205). As for Hypothesis 2, Newton rejected a dense, Cartesian aether from the time of the composition of the *Principia*, as for example explicitly in *Opticks*, Query 28. The last sentences that he drafted on this topic demonstrate a strong preference for branch 2(b). Further, there can surely be no room for doubt that this 'electric and elastic aetherial medium' (to combine the words of *Principia* and *Opticks*) was a *mechanical* aether, not a spiritual medium, extended but immaterial. For it is particulate, Newton makes clear; it also possesses density and elasticity, properties which Newton is able to compare numerically to the similar properties of air; how could such properties belong to or be necessary in a spiritual medium? More's conception of a Spirit of Nature and Newton's of a universal aetherial medium operating mechanically on bodies are surely utterly different and there can be no intellectual filiation between one and the other.

But what of the 'active principles' of which Newton also wrote, and which so many historians have associated with his reading of Henry More? It is fairly obvious that these might be worked into versions of either Hypothesis 1 or Hypothesis 2, though Newton does not seem to have done so. It is certain that he wrote of both an 'active principle' and the action of a rare aether in relation to gravitation and cohesion (at least), so the one does not exclude the other. It is possible that he would have regarded 'active principle' as *cause* and aether as *agent*. We can but guess. However, historians can be more confident about the temporal pattern of Newton's changes of mind about atoms and void, principles and aether. They began with a phase of aetherial speculation (manifest in his *Hypothesis explaining the Properties of Light* (1675) and *Letter to Boyle* (1679), neither printed in Newton's lifetime[11]), which had ended before Newton began to compose the *Principia* in 1684. Probably his view of things changed when he plunged into mechanics and the study of comets in 1679–81, for at this time he finally renounced the Cartesian postulation of vortical rotation as the agent of the planetary revolutions, and began to consider seriously the idea of a universal gravitational force associated (if only phenomenalistically) with matter. The whole writing of the *Principia* was based on the pre-supposition of atoms in the void as the ultimate reality of physical Nature. The same presupposition dominates the *Opticks* (1704) and its Queries; Newton argued strongly in favour of the emission or material theory of light and against the undulatory theory (which of course required a medium of transmission, or aether) and also against the supposition that the celestial spaces could be filled with such a medium. It was in these years too that Newton wrote elaborately about the important functions in Nature of the 'active principles', especially in the extra Queries added to the Latin *Optice* of 1706.

Thus we find that (in these respects) the influence of Henry More upon Newton appears strongest not in his early manhood, but in his mature years

of greatest achievement. It was in these years that he was most drawn to
the notions of the emptiness of space, across which forces operated deriving
from 'active principles', and through which the streams of light-particles
penetrated. Such ideas enabled dynamics to take shape as an abstract
mathematical science undisturbed by physical explanations of the causation
of forces. At this time too Newton at last published his ambition to extend
the idea of force to chemistry.

By 1713, pressed by the wish to escape the taunts of Leibniz and the
need to give some physical meaning to the concept of gravitation, impressed
by recent laboratory demonstrations of phenomena seemingly requiring
the existence of physical media or aethers for their explanation, Newton
once more recoiled from More and invoked the physical and physiological
'spirit' of the last paragraph of the second edition of the *Principia* (Guerlac
1977: 107–28). Four years later, the second English edition of *Opticks*
committed him more deeply and irrevocably to the notion of an exceedingly
rare, highly elastic aether which is also the electric and elastic spirit, while
(confusingly) leaving the new material of 1706 on 'active Principles' still
largely unchanged. The alteration in Newton's thought may be illustrated
by comparing these sentences of Query 20 in 1706 with their equivalent
in Query 28 of 1717:

> 1706   Moreover, that there are absolutely no fluid mediums of this kind I
> gather from the fact that the planets and comets are borne with
> such regular and enduring motions through the celestial spaces,
> everywhere and in all places and in all directions. For thence it is
> manifest, that the celestial spaces are devoid of all sensible resistance
> and so of all sensible matter. (1706: 310, my translation)[12]
>
> 1717   And against filling the Heavens with fluid Mediums, *unless they be
> exceedingly rare*, a great Objection arises from the regular and very
> lasting Motions of the Planets and Comets in all manner of Courses
> through the Heavens. For thence it is manifest ... [*as before; italics
> added*]. (1952: 365)[13]

It is evident that Newton has weakened his former argument in order to
accommodate his new one. It cannot be said that Newton's desertion of
More and return to an idiosyncratic neo-Cartesianism improved the internal
consistency of his declarations upon the more obscure topics in natural

philosophy, or enhanced his reputation as a metaphysician.

From these textual considerations one sharp point emerges: if Newton found More's metaphysics useful in enabling him to proceed upon dynamic rather than kinematic principles in mechanics, he nevertheless felt himself perfectly free to abandon them later. For, whether or not we choose to regard Newton's 'active Principles' as consistent with the 'rare aether' introduced in 1713, it is certain that his idea of space, at least, has suffered a modification. It is no longer a void (save where matter is present) in the physical sense, though of course filled with immaterial spirit: this was More's conception and then Newton's. But Newton's new idea of space is that it is charged with a highly active aether. Whether or not 'active Principles' are still to be postulated as the cause of forces in Nature, Newton now says that this rare aether is the means by which they operate *mechanically* upon matter. The immateriality of the means by which 'active Principles' work upon matter, as forces, has been surrendered and with it the former clear distinction between matter as inert and non-matter as active. God has certainly, even after these changes, an essential rôle in the universe – this, after all, was the burden of the General Scholium in which Newton introduced the rare aether for the first time – but He (or His surrogate, the Spirit of Nature) is no longer directly implicated in the phenomena of light and cohesion, attractive force, chemistry and nerve action.

Or at least it would be so had Newton made a decisive choice. But as many have pointed out, and their form permits, the gradually assembled Queries present many possible views at once, the later accretions not wholly excluding the earlier thoughts: 'none of [the] explanatory models is ever repudiated by [Newton]; all continue to play a part in his shaping of the ontological alternatives' (McMullin 1978: 79). Historical attempts to reconcile these various elements in Newton's thoughts, or even to discover a clear pattern in them, have so far failed. One might suppose that he was content to offer different solutions to suit the prejudices of different readers, or the various possibilities of future discovery. All that is significant, in relation to Henry More, is that Newton seems to have moved away from his Platonist ideas. As McMullin puts it: 'Newton's strong belief in the passivity of matter was rooted in the older natural philosophies, particularly neo-Platonism. The new mechanics seemed to be leading him, almost inexorably, to a modification of this principle. Yet he could never quite bring himself to accept this' (1978: 108). Rather as Einstein, in a situation in the history of physics not wholly unlike Newton's, rebelled against the acceptance of quantum mechanics. Both felt the power of indisputable mathematical logic; yet both felt strong ties to a metaphysical conception of reality.

One has to take a very generous measure of More's influence upon Newton to discern in it any profound effect upon his scientific achievements, and even so far as Newton's metaphysical pronouncements are concerned it is easy to exaggerate More's effect. A preference for atoms and the void Newton could as well have drawn from his reading of Walter Charleton; his ideas of space and time were possibly as much shaped by Isaac Barrow as by More; his criticisms of Descartes, though owing something to More's books, were seemingly unaffected by More's celebrated letters to the French philosopher. Finally, as Newton in late maturity and old age pondered the problems of cosmology and ontology, he became less and less satisfied with straightforward spiritualist solutions of them.

Having said this, the historian might wish to add as a tailpiece that in the light of hindsight the overall theory produced by Newton's reading of More is far more interesting than is his return to aetherial mechanics. Although theories of 'subtle fluids' and aethers were enthusiastically canvassed in the eighteenth century by many philosophers, with some indebtedness to Newton, and although (under the stimulation of the vindication of the wave-theory of light, which again owed some debt to Newton at least in the case of Thomas Young) many nineteenth-century physicists believed that the existence of an aether was as certain and essential as that of matter itself, the concept of the aether proved an *impasse*. The more elaborate the attempt to perfect it, whether by Bryan Robinson or Joseph Larmor, the more conspicuous the failure, mathematical sophistication barely concealing the frustration of basic ideas. Given the conviction from which Newton never departed, that God is both the creator and the perpetual upholder of the universe, it was incumbent upon the philosopher to make his belief valid in philosophy. So Newton opened his correspondence with Richard Bentley with the famous words, as solemn as they are appropriate: 'When I wrote our Treatise about our System, I had an Eye upon such Principles as might work with considering Men, for the Belief of a Deity, and nothing can rejoice me more than to find it useful for that Purpose' (Cohen 1958: 280). It was no more outrageous for a philosopher to maintain that the omnipresent divine spirit, filling space, is the ultimate source of the active principles in Nature, than for a theologian to assert that God's will rather than necessity is the cause of all things. Such ideas at least did not restrict the philosopher by arbitrary hypotheses, however ingenious: More's notions, as interpreted by Newton, paradoxically made for intellectual freedom.

Further, it is surely clear that the spiritualist view of Nature entertained by the Cambridge Platonists offered Newton a better ground for explaining the place of God in natural philosophy than did his revised aetherial theory. Not, of course, the Christian God of More and Cudworth for, as Westfall has recently re-emphasized (Lindberg and Numbers 1986: 228–35), Newton took a long step on the road to establishing deism as the only rational form of belief. The best way to assert the conviction that God continually and actively upholds our universe is by specifying the functions that He performs – functions properly associated with His omnipresence, intangibility and willingness to submit His omnipotence to the laws and conditions by which He has Himself defined that universe. The analogy that Newton suggested (but was too timid to embrace wholeheartedly) between the human soul in the body and God in the spatial infinity of the universe is a very elegant one, for there is just that same element of unfathomable mystery in the action of spirit upon matter in humanity, as in the action of spirit upon matter in the universe at large. By making some concession to the mechanistic, unmiraculous philosophy of Leibniz, Newton not only spoiled the analogy and destroyed the mystery but opened the door to the idea that even creation demands no divine power.

APPENDIX I

The Chief Philosophical Writings of Henry More

1642 *Psychodia Platonica: or a Platonicall Song of the Soul, consisting of foure severall poems*

1646 *Democritus Platonissans*

1647 *Philosophical Poems* (revised and enlarged texts)

1650 *Observations upon Anthroposophia Theomagica and Anima Magica Abscondita*

1653 *An Antidote against Atheisme*

1653 *Conjectura Cabbalistica*

1656 *Enthusiasmus Triumphatus*

1659 *The Immortality of the Soul*

1660 *An Explanation of the Grand Mystery of Godliness*

1662 *A Collection of several Philosophical Writings* (so-called second edition)

1664 *A modest Enquiry into the mystery of Iniquity* and *Apology of Dr. Henry More*

1664 *Epistola ad V.C.* (separate issue)

1668 *Enchiridion Ethicum*

1668 *Divine Dialogues*

1671 *Enchiridion Metaphysicum*

1676 *Remarks upon the two late ingenious Discourses*

1679 *Ad V.C. Epistola altera* in *Opera Omnia.*

The *Collection of Philosophical Writings* reprints revised texts of the *Antidote against Atheisme, Conjectura Cabbalistica, Enthusiasmus Triumphatus* and *The Immortality of the Soul.* I suppose this accounts for the words 'Second Edition' on the title page – for there had been no earlier issue of this folio volume. However, the correspondence with Descartes and the *Epistola ad V.C.* also included in it had not been previously published by More.

A full bibliography of More, compiled by Robert Crocker, concludes Hutton 1990.

More and Galileo, 1647

It has perhaps not been sufficiently appreciated that Henry More may have some claim to have played a part in disseminating the work of Galileo in England. The *Notes* to the Platonical Songs were first published in their second, enlarged edition of 1647, that is, at roughly the same time as the better-known popular books of John Wilkins. The *Notes* explain in some technical detail the scientific allusions in the texts of the poems, some of these being first printed in 1642.

It may be doubted that More read *De revolutionibus orbium coelestium,* and confidently surmised that he drew his knowledge of the Copernican system from Galileo's *Dialogo* (1632) and other sources. On p. 379 of the *Philosophical Poems* (1647) he gives the following dimensions within the solar system, taken from 'Landsberg' – perhaps the *Progymnasmatum astronomiae restitutae* (1619) of Philip van Lansberge:

Distance of the Moon from the Earth at apogee 64 Earth-radii,
at perigee 54 „ „
Distance of the Sun from the Earth at apogee 1550 „ „
at perigee 1446 „ „

On p. 388 he gives the following distances from the Earth, in Earth-radii:

Mars at perigee	556	Venus at apogee	2598
Venus at perigee	399	Mercury at apogee	2176

On pp. 389 and 390 are geometrical representations of the Tychonic and Copernican systems, respectively. More remarks: 'It is plain to any man that this

[Copernican] System of the World is more naturall & genuine than that of *Tycho's*. No enterfaring or cutting of circles as in *Tycho's* . . .'

A little later in the *Notes* (pp. 400–1) More explains with a diagram what is meant by the phases of Venus, and how they furnish evidence for the revolution of this planet (and Mercury) around the Sun. Immediately afterwards he continues with an account of the stations and retrogressive motions of the planets, indicating the superiority of the Copernican treatment, though – as is clear from his diagrams – this is a little more difficult to understand than the representation by an epicycle.

In annotations to a short, natural-theological poem entitled *The Philosopher's Devotion*, More explains how the daily alternations and seasonal variations of the length of day and night are brought about, on the assumption of the Earth's daily rotation on its axis and annual revolution about the Sun. More writes that he follows '*Copernicus* his *Hypothesis* [which] will not merely explane these verses but exceedingly set out the fitnesse and genuinenesse of the Hypothesis it self. Which I will therefore do out of *Galileo* for the satisfaction of the unprejudiced and ingenuous Reader.' Once again, More repeats Galileo's diagrams.

A very interesting section of the *Notes* (pp. 390–400) deals in some detail with the theory of the tides. The original verses of *Psychathanasia* (Book III, Canto 3, stanza 56) adopted the Galilean kinematic explanation without question, and More first explains what this involves (pp. 391–5): the flux and reflux of the sea are due to diurnal and seasonal changes of the speed through space of any point on the Earth's surface. However, on p. 396 More affirms that the Galilean theory must be wrong, because it makes the annual variations in tidal movement greatest at the solstices, whereas they are found in fact to be greatest at the equinoxes. Hence he now declares Descartes to be 'far more Successfull in his Hypothesis' of aetherial pressure as the cause of the tides, and this More summarizes from *Principia Philosophiae* (1644).

Two other points treated by More in the *Notes* and relating to physics are worth a mention here. One is a very poor discussion of the well-known problem of the vertical rise and fall of heavy bodies (p. 385). The other is a note (p. 425) defining *Circulation* (which, it is suggested, might better be called *orbiculation*), a term applied to the diffusion of an effect from the centre as when rings expanding outwards are formed on the surface of water by the splash of a stone: 'In brief, any thing is said to circulate that diffuseth its Image or Species in a round . . . Such is the diffusion of the Species audible in the strucken Air, as also of the visible Species.'

More's Books and the Fellows of the Royal Society

I examined the catalogues of seven large libraries owned by Fellows of the Royal Society in the late seventeenth century: those of Elias Ashmole, Robert Hooke, Edmond Halley, John Ray, Henry Oldenburg, Isaac Newton and Isaac Barrow.[1] Among the tens of thousands of volumes dealing with every learned topic – and some less learned ones – I found recorded twenty-eight philosophical volumes of which More was the author. In addition there were eight books by More dealing with devotion and theology; these eight belonged to either Barrow or Newton. Newton and Ray owned the largest number of More's works: nine each. Halley owned the least number, uniquely the *Immortality of the Soul*. Whether possession of this single work confirms or weakens the story that Halley was a freethinker others must judge. None of the seven scientists seems to have owned the folio *Collection of Philosophical Writings* of 1662.

The frequency of More's philosophical works in these seven libraries is as follows (titles abbreviated):

Antidote (1653)	5
Immortality (1659)	4
Philosophical Poems (1647)	3
Conjectura Cabbalistica (1653)	3
Enchiridion Ethicum (1668)	3
Enchiridion Metaphysicum (1671)	3
Divine Dialogues (1668)	2

Remarks (1676)	2
Observations (1650)	1
Epistola ad V.C. (1664)	1
Letters (1694)	1
	28

Because More was such a prolific author, one might expect his later works to have been printed in large editions, and several were popular enough to be reprinted more than once. But my small sample indicates that among the scientists More's early works were the more frequently acquired, and that his books were not very popular among the FFRS. One could easily cite examples of other books that, like Ralph Cudworth's *True Intellectual System*, were found in all seven libraries. Moreover, Newton's collection certainly and Barrow's and Ray's possibly were enlarged by presentation copies. Though only Halley's library indicates a definite lack of interest in More's books, the collections of Ashmole and Hooke only contained three titles each, a rather slight representation. I suspect that More appealed far more to 'non-scientific' than to 'scientific' readers, anachronistic as it may be to apply such a dichotomy to More's time.

Notes

CHAPTER 1 *Introduction*

1   More than a century ago portraits of the Founders, principal early benefactors and great Christ's men were inserted into the west oriel window of the College Hall.
2   A phrase from Ralph Cudworth's (printed) 1647 sermon before the House of Commons.
3   Henry Burrell left an annual rent-charge of £20 to the College, of which £4 was to be spent on the purchase of 'new books and mathematical instruments'. Henry More's tutor, William Chappell, was executor of Burrell's will. An MS list of early book purchases survives.

CHAPTER 2 *Platonism*

1   The name of Gaius Caesar (Caligula) in Greek characters adds up to 616; Nero Caesar, 616 likewise; Neron Caesar, 666. Both 616 and 666 have been asserted to be the number of the Beast (that is, the Roman Empire) in the *Book of the Revelations of St John*.

CHAPTER 3 *Platonism and the Scientific Revolution*

1   McGuire and Rattansi 1966: 133. The *locus classicus* for the development of this opinion is Theophilus Gale, *The Court of the Gentiles*, 3 vols, 1669–77.
2   A *locus classicus* is Randall 1940: 177ff. A great deal more scholarship has been devoted to this topic in recent years by the late Charles B. Schmitt.

3   Jan Marcus Marci of Prague anticipated Newton in the discovery of the dispersion of a ray of light refracted through a prism, but this discovery did not produce for Marci, as it did for Newton, immensely fertile theoretical consequences. See his *Thaumantias*, Prague, 1648.

CHAPTER 4   *The Cambridge Platonists*

1   For a longer Scholium on ancient atomism drafted for the *Principia* see Casini 1984: 36–8. See also Copenhauer 1980.
2   This *Query* was first published in the Latin *Optice* (1706) where it is numbered 20; the adjective *dense* was added to the next English edition (1717). See Guerlac 1963a: 29–31.
3   Newton's scholarly researches, which depended heavily upon the *Mythologiae libri decem* (Venice, 1551) of Natale Conti, were some years posterior to the writing of the *Principia* and are irrelevant to its genesis. Newton's purpose in pursuing them was to summon ancient endorsement of the concept of universal gravitation against his critics who called such a concept absurd; see Casini (1984). Newton's annotations upon Cudworth's *True Intellectual System* are now in the Clark Library, University of California, Los Angeles; see Westfall 1980: 353; 510; and McGuire and Rattansi 1966.

CHAPTER 5   *Henry More, Man of Paradox*

1   Cf. Henry More's Will (Nicolson 1930: 482–3) naming Gabriel More, his brother Alexander's son, as his executor and residuary legatee.
2   Richard Ward (1658–1723), More's biographer, was born at Sheering in Essex. He became a sizar at Christ's College, Cambridge, and may therefore have been one of the unfortunate amanuenses of whom More so often complained. He proceeded MA in 1681, and was Rector of Ingoldsby, Lincs., from 1685 onwards.
3   The *Theologica Germanica* was an anonymous work of the fourteenth century, reissued by Luther in 1516.
4   Perhaps the verb concluding line 3 should be 'fit'.
5   The final blow of fate to Worthington was the destruction of his London Church in the Great Fire of 1666.
6   Standish was a Fellow from 1653 to 1714. To these names Nicolson (1929a: 40) adds those of Thomas Fuller and Joseph Sedgwick, who both came to Christ's from Catherine Hall.
7   The book was *Conjectura Cabbalistica ... Or a Conjectural Essay of interpreting the Mind of Moses according to a threefold Cabbala ...* (1653b). Cudworth, 'accomplished in all parts of Learning ... as well in the *Oriental* tongues and *History*, as in all the choicest Kinds of *Philosophy*' is asked to accept the dedication of the book 'as a Monument or Remembrance of our mutual Friendship'.

Notes

8 See Pepys 1974: 21 February 1659/60. Nicolson (1929a) examines the Widdrington affair in detail.

9 More continues: 'For I understand not the looseness of their language, nor whom they mean, nor what they would have.'

10 A belief – really essential to the idea that the Mind is not a *tabula rasa* at birth – also taught by More's pupil George Rust in *A Letter of Resolution concerning Origen*, 1660.

11 It appears that in 1649 or 1650 Anne Finch wrote to More for assistance in her study of Descartes's philosophy; More's first reply is dated 21 February [1650] (1930: 51). She was then aged 18. When, in 1652, More dedicated to Lady Conway (as she had become) *An Antidote against Atheisme*, he wrote, her 'Genius I know to be so speculative, and Wit so penetrant, that in the knowledge of all things as well Natural as Divine, you have not onely outgone all your own Sex, but even of that other also, whose ages have not given them over-much the start of you.' The lady naturally blushed at the compliment (1930: 71).

There is a brief life of Anne Conway in Fraser 1984: esp. 390–400.

12 Miss Nicolson accepted More's idealized view of Finch and Baines. For the opposite, see Abbott 1920. The portrait of neither man (Fitzwilliam Museum, Cambridge) suggests a happy disposition.

13 More wrote to Lady Conway (? Spring 1660): 'I should as naturally come and wayt upon your Ladiship as the needle turns to the North, but that I have not dispatched all my businesses here … in favour and satisfaction to myself, who enjoy so much contentment in your Ladiships converse' (1930: 162, cf. also 271).

14 The dinner had taken place the day before.

15 As late as 1674 More again alludes to 'my wonted exercize of running about the Orchard after Chappell' (p. 399).

16 The writer was then only 23 years old.

17 In 1685 Miss Cudworth had married, as his second wife, Sir Francis Masham, whose country house was at Oates in Essex; here Locke spent the last years of his life.

18 This letter was misattributed by Nicolson as a letter of 2.vi.1653.

19 Ezekiel Foxcroft was related to two Masters of Emmanuel College: Benjamin Whichcote (his mother's brother) and John Worthington.

20 Now known as the Fellows' Building, built during the Civil War. Both More's father and his uncle Gabriel contributed to its cost. More lived in the north-east chamber on the second floor, I take it on what is now 'A' staircase.

21 According to Ward (1710: 174) More found 'the great mystery of godliness … delineated in the Visions and Prophecies of Scripture, and particularly those of Daniel and the Apocalypse'.

22 Finch outlived his friend by only a few weeks; both were major benefactors of the College.

CHAPTER 6 *Henry More's Philosophy*

1   Modern experts on the Hebrew Cabbala do not regard More's knowledge of it as deep – nor did he.

2   The subtitle of Gale's work in three stout volumes is *A Discourse touching the Original of* HUMAN LITERATURE, *both Philologic and Philosophic, from the* SCRIPTURES *and* JEWISH CHURCH. Gale's object is to demonstrate the perfection of God's word, the imperfections of Nature's light, and the right use of sound philosophy. Like More, Gale cited ancient testimony to the fact that Pythagoras and Plato had taken their philosophy from the Jews. It was 'Plato's usual way (as Pythagoras before him) to wrap up those *Jewish Traditions* in *Fables* and *enigmatic Parables*'. Further, 'all *Languages* and *Letters* had their *derivation* from the *Hebrew*.' Gale again agreed with More that '*Pagan Physicks*, or *Natural Philosophie* ... seems evidently traduced from the first chapter of *Genesis*, and some *Physick Contemplations* of Job,' via the hands of Mochus and Thales; mathematics also comes 'from the *Church of God*'. Perhaps too it was the Children of Israel's first division of the land of Canaan that gave birth to geometry. After quoting Steuco at some length on the derivation of the perennial philosophy from the ancient Hebrews, Gale finds the same idea strongly supported by Protestant scholars – from Melancthon through both Scaligers and the elder Vossius to the contemporary English, including Cudworth.

3   The *Epistola ad V.C.* follows immediately after More's reply (written long after Descartes's death) to the latter's final fragmentary reply to his letters (at pp. 108–33).

4   *Ibid.* Such ideas, especially the likening of the 'Spirit of the World' to what would later be called the 'vital principle' in living things, were old with More; cf. the *Notes* to *Philosophical Poems* (1647: 345): '*Physis* is nothing else but the vegetable World, the Universall comprehension of Spermaticall life displayed throughout ... not the divine Understanding itself, but an Artificer's imagination separate from the Artificer, and left alone to work by it self without animadversion.'

5   The 'Rules' appear in *The Apology of Dr Henry More*, published with *A Modest Enquiry* (1664b).

6   Compare Lord Conway to his daughter-in-law Anne: 'Learning came to the Greeks from Aegipt, from Phoenicia, from the Jewes, and from the Indies. From Greece it came into Italy, from thence into France, Germany and England. The Druides were the learnedest sort of men which France and England had but they had no more learning than a heathen West Indian priest hath, which is very small' (Nicolson 1930: 35, 20/9/1651).

7   In his print (1762) 'Credulity, Superstition, Fanaticism'.

CHAPTER 7 *The Spirit World*

1   Walker (1986) emphasizes the relation between More's concept of spirit and the spirits of physiology. I confess inability to see the force of this argument.

2   See Nicolson 1930: 423 for More's concurrence with Sir Mathew Hale in 'exploding that monstrous spring of the ayre' (February 1676).

3   Though Finch and Baines did not leave England for another twelve months, there was no further meeting with More. Finch kissed the King's hand for his appointment, but he was the nominee of the Levant Company, which paid his salary.

4   For earlier stories see e.g. Nicolson 1930: 270, 294 ('... if I can gett sure instructions, I will tell the Story of the Spirit of a murdered Boy in Sheffield ...' – 12/5/1668).

5   The Sadduccees were an ancient Jewish sect denying the resurrection of the body, personal immortality, etc. The full title of the 1688 edition of the book is *Saducismus Triumphatus: or Full and Plain Evidence concerning* WITCHES *and* APPARITIONS *In Two* PARTS ... The title of the 1666 version was *Philosophical Considerations touching Witches and Witchcraft*; reprinted as *Philosophical Considerations against Modern Sadducism in the matter of Witches and Apparitions* in Essay 6 of Glanvill 1676. There are other versions besides. The text takes in later editions the form of a letter to Robert Hunt JP.

CHAPTER 8   *More and Descartes*

1   Probably More would not have shared Locke's view that a young gentleman should study natural philosophy for the sake of finding topics for polite conversation.

2   So Nicolson 1929b: 362, but the expression is qualified a page or two later. Miss Nicolson was perfectly aware of the critical force of More's 1648–9 letters.

3   1664a: 42: 'culpandus sit ... quod Mathematico suo Genio ac Mechanico in Phaenomenis Naturae explicandis nimium quantum indulserit.'

4   See Descartes 1974: 235–6; 246–50; 629–42. Clerselier obtained from More 'transcripts' of the letters he had written to Descartes in 1655; More had revised, improved and shortened the originals. He republished them himself in *Philosophical Writings* (1662). See also Webster (1969).

5   This passage is translated in Burtt 1932: 131–2.

6   More's own words about Gassendi are coarse: how could he have the patience to rake old rags out of the rotten dunghill of Epicureanism to stuff his large volumes with?

7   I prefer to read these letters in their order in the MSS, which seems to me to make better sense than Miss Nicolson's rearrangement. The letters are dated without year, and could conceivably belong to 1648–9 rather than 1649–50. Moreover, the letter of February shows that More did not initiate the correspondence, while that of August acknowledges a (presumably first) letter from Anne Finch brought to More by her brother.

8   9 September [1650]. More's philosophical letters to Anne Conway, omitted by Miss Nicolson, are printed in Gabbey 1977 with analytical notes.

9   *Dialogues* 4 and 5 were first published; the interlocutors are not the same as those of *Dialogues* 1 to 3, published later.

10 Miss Nicolson (1930: 216n) proposed that Philotheus was modelled on Ralph

Cudworth, Hylobares on the materialistic Thomas Baines. But, as the quotation shows, such identifications are fanciful enough.

11 Descartes shared this conception. He also held that matter is indifferent equally to motion and to rest; in the absence of external impulse or impediment a particle remains either at rest or in uniform rectilinear motion. More seems (like Kepler) to believe that matter is naturally sluggish and resistant to motion.

12 Laplace would have disagreed!

13 'Models' is of course a modern term.

14 Alan Gabbey puts the open alienation of More from Descartes to about 1655.

CHAPTER 9 *More and the Royal Society*

1 Laurence Rooke (1622–62) was Gresham Professor of Geometry.

2 The First Charter of Incorporation was dated 15 July 1662.

3 Worthington reports that More's works were very imperfectly printed due to his tiny handwriting and careless correction of the press, but it was also the case that More deliberately chose archaic or distorted forms of words; one has often to guess his meaning.

4 On More's avoidance of Descartes's moral philosophy and metaphysics see Gabbey 1982: 186; 201.

5 On Hartlib, see Webster 1969 and Webster 1975. He was not FRS.

6 Another classical 'Platonic' hallucination: 'If after a decoction of *hearbs* in a Winter-night, we expose the liquor to the frigid air; we may observe in the morning under a crust of Ice, the perfect appearance both in *figure* and *colour*, of the *Plants* that were taken from it' (1661: 46).

7 The attacks upon and defence of the Royal Society have been much studied; see e.g. Syfret 1950.

8 More's letter was printed in Glanvill 1671: 153–7.

9 cf. 1671: cap. 12, where More rehearses his 'refutation' elaborately with diagrams. He had a similar difficulty, as noted above, in understanding how a circular disk of wood, closely (but not tightly) fitting a cylindrical bucket, can float to the surface when the bucket is filled with water. Why does not the weight of superincumbent water hold the disk down, since the area of the disk is much greater than the area of the gap between the disk and the bucket?

10 The publication of Blaise Pascal's *Traitez de l'Equilibre des Liqueurs et de la Pesanteur de la masse de l'Air* was delayed until 1663. The experiments of the Florentine Academia del Cimento (1657–67) were still unpublished.

11 The question at issue is not uniquely one of Boyle's own experiments, as his allusions to Archimedes indicate.

12 The letter should be read as written from Christ's College, Cambridge. (C.C.C.). In the transcript in Nicolson (1930: 264) the name printed 'Fulwood' is inexplicably rendered as 'Fleetwood'.

13 The writing was finished in March 1667.

14 More had not yet seen Boyle's *Hydrostatical Discourse.*

15 The letter is simply dated C[hrist's] C[ollege] C[ambridge], Dec. 4, but the year must be 1672 because Boyle's *Discourse* was probably not written by December 1671 and certainly More had not read it by then. The publication of the *Discourse* was not before 13 February 1673, the date of the following *Hydrostatical Letter* addressed to George Sinclair.

16 Westfall (1971: 207) suggests that 'force' here is equivalent to the later concept of 'work' while stressing the total confusion in Hooke's ideas.

17 I find it difficult to believe that Hooke had a strong sense of the Christian religion or that the discovery of God in Nature was of profound importance to him. There is no evidence of private piety in his diaries. But *Micrographia* contains public manifestations of his belief in the divine design and benevolence.

18 My belief is that More used abbreviations for *Virum clarissimum* (or *celeberrimum*), the ordinary polite form of address for a letter.

19 Marion (1880: 12) points to the concept of irritability as already present in *Anatomia hepatis*, 1654.

20 The *Tractatus* is apparently a very scarce work.

21 Marion 1880: 32 – 'vitam spirituum esse substantialem', 36.

CHAPTER 10 *More and Newton: Space and Time*

1 The leading secondary authorities are Burtt 1932 and Koyré 1957.

2 The subtitle is 'Four chapters in the Explanation of the grand mystery of Godliness, which contain a brief but solid confutation of judicial astrology.' The *Grand Mystery of Godliness* was first published in 1660.

3 I infer that the inscriptions recording the gifts are in Newton's hand.

4 Koyré also quotes (pp. 153–4) a single passage from *Enchiridion metaphysicum* to show that More reverted to the use of Descartes's word *indefinite* for the universe so reserving the term *infinite* for God.

5 Of course, seventeenth-century clocks were not more uniform in their motions than the heavenly bodies, but they were accurate enough to reveal that 'solar time' is not uniform throughout the year. Huygens had found it necessary to correct it seasonally by the 'equation of time' to arrive at *mean* solar time. Probably Barrow had this in mind.

6 Leyden notes (1968: 236–8) that J. B. van Helmont (whom Newton read) expounded a notion of absolute time.

7 It is unfortunate that *De gravitatione* cannot be dated with precision. It is very unlikely to have been composed before 1664, the date of Barrow's lectures, and it most probably belongs to the period when Barrow and Newton were closely associated.

8 It is strange that Newton does not employ the notion of potentiality here.

9 Newton made various alterations in the third edition of the *Principia* (1726) without changing the sense.

10 Literally, 'will not be *nowhere* and *never*'.

11 Leyden (1968: 256, n. 6) records the connection between More and Newton on absolute dimensions noted by Joseph Raphson, a rather obscure Newtonian mathematician, in 1702.

Chapter 11 *More and Newton: Force*

1 Westfall (1971) seems to have overlooked the possible relevance to Newton of More's criticism of Descartes.

2 I have had to guess the meaning of *substentaretur; ?* read *substantiaretur*.

3 In subsequent editions Newton adopted forms of words echoing More's ideas less closely, without altering the basic meaning.

4 See Westfall 1971: 338 and Westfall 1975: 218 expressing a view similar to that advanced here.

5 The first two Rules of Reasoning are called *Hypotheses* in the first edition. Perhaps it should be noted here that Newton extended More's (or Epicuros's) notion of the material emptiness of space between substantial bodies (if the tautology may be forgiven!) to the interior of substance itself; that is, he argued that although we find some substances to be hard, rigid, massive and resistant to fracture, nevertheless even these are composed largely of void, within which the solid atoms are widely interspersed. It is thus readily possible for short-range forces to operate in the internal spaces between the ultimate atoms (or higher orders of particles), just as long-range forces operate in the greater spaces between gross bodies. The *locus classicus* is *Opticks*, Book II, Part 3, Proposition VIII (Newton 1952: 268–9), a passage virtually translated from that first published in Latin as a *Corrigendum* to *Optice* (1706), *ad init.* Cf. Thackray 1970: 53–67.

6 Newton's comment about the non-deliquescence of common salt argues a certain want of domestic experience!

7 Modern philosophers criticize Newton's views (in the Third Rule of Reasoning) about the inference of the properties of fundamental particles, such as impenetrability, from the same property in macroscopic bodies. Cf. McMullin 1978: 13ff. NB, here Newton's repulsive force exactly replaces More's 'Spirit of Nature'.

8 My interpretation is shared by McGuire (1977: 107).

9 The final English version of the passage discussed is in Newton (1952: 403): the wondrous effects of Creation 'can be the effect of nothing else than the Wisdom and Skill of a powerful ever-living Agent, who being in all Places, is more able by his Will to move the Bodies within his boundless uniform Sensorium, and thereby to form and reform the Parts of the Universe, than we are by our Will to move the Parts of our own Bodies. And yet we are not to consider the World as the Body of God, or the several Parts thereof, as the Parts of God ...'

10 The date of Newton's letter to Bentley is 28 February 1693; the letters were first published in 1756.

11 However, Bentley also writes (p. 344) that gravitation is 'above all Mechanism and material Causes, and proceeds from a higher Principle, a Divine energy and impression'.

12 Newton reverts to a modified form of earlier speculations.

13 I do not agree with Hesse (1961: 152) that: 'It is clear that Newton thought there must be some physical cause of gravity still to be found . . .', for it seems to me that Newton's own arguments against such a physical cause (understood in seventeenth-century terms) are overwhelming. And (as Professor Hesse indicates at the end of her paragraph) there was no future in looking for such causes.

CHAPTER 12 *Conclusion*

1 Ray refers to More 1659. Apparently he took nothing from More's later books, though he owned them (see appendix III).

2 Spinoza was supposed by many, More among them, to have embraced a complete materialism, an accusation that Spinoza himself rejected.

3 How far the 'as if' stretches in the mind of Spinoza is indicated in the following sentence: '[W]hat we get from Spinoza is a physical theory (or theory-sketch) that is neither drawn *from* nor applicable *to* a concrete embodiment in particular natural phenomena' (Nancy Maull in Grene and Nails 1986: 12, italics in original).

4 See chapter 11, note 5. A sphere fills rather more than half the volume of the circumscribed cube. So if spheres are packed in rows and columns, touching each other, they occupy a little more than half the available volume.

5 This passage provoked the contempt of Leibniz who accused Newton of supposing that the universe is God's nervous system. It has been suggested that Leibniz read one of those copies of the Latin *Optice* (1706) in which the word *tanquam* ('as it were') does not appear: Cohen and Koyré 1961.

6 See Curtis 1959; Hill 1965; Feingold 1984. More has not been a major figure in the debate.

7 'Sandalphon' is the name of one of three Hebrew angels, who receive the prayers of the people and weave them into garlands.

8 There is no single reference to Henry More in D. Brewster, *Memoirs of Sir Isaac Newton* (1855).

9 The note records other French writers of the period who (in addition to A. E. Burtt and A. J. Snow) took a similar view. Meyerson – my translation – gives within the quotation marks a French summary of the last paragraph of the *Principia* (1713) on the 'electric and elastic spirit'.

10 These four words are to be found only in Motte's English version of the *Principia*, but decisive evidence justifies them as stemming from Newton (see Guerlac 1977: 128, n. 4).

11 See Newton 1959–77, I: 362–86; II: 288–95. Both documents were first printed in the eighteenth century.

12 McMullin (1978: 95–6) takes the aether of 1717 to be less an extension of the concept of the 'spirit' of 1713 than a new hypothesis altogether. Like Guerlac (1977: 127) I find it simpler and more natural to consider 'spirit' as a prototype of 'aether'. I certainly can see no great significance in the change of name – unless this was a sop to the Cartesians.

13 Newton added new Queries numbered 17 to 24 to the 1717 edition of *Opticks* which are largely concerned with the rare, immensely elastic aether. Consequently, the Queries numbered 17 to 23 in the 1706 *Optice* were now renumbered (in slightly revised form) as Queries 25 to 31.

Appendix III *More's Books and the Fellows of the Royal Society*

1 For the first four see Feisenberger 1975; the dates of the sales were respectively 1694, 1703, 1742 and 1708. Oldenburg's library was bought entire by the Earl of Anglesey and is therefore represented in *Bibliotheca Angleseiana, sive CATALOGUS variorum librorum* ... (1686); for Newton's library see Harrison 1978, and for Barrow's, Bodleian Library MS Rawlinson D878, f.39 ('A Catalogue of the bookes of Dr Isaac Barrow sent to S.S. by Mr Isaac Newton Fellow of Trin: Coll: Camb: July.14. 1677').

Bibliography

If none other is stated, the place of publication is London.

Abbott, G. F. (1920) *Under the Turk in Constantinople, 1674–81.*

Aiton, E. (1985). *Leibniz: A Biography*, Bristol and Boston.

Allen, M. J. B. (1975) *Marsilio Ficino: The Philebus Commentary*, Berkeley.

Armistead, J. M. (1986) 'The Occultism of Dryden's "American" Plays in Context', *The Seventeenth Century*, 1(2), 127–52.

Axtell, J. L. (1968) *The Educational Writings of John Locke*, Cambridge.

Ayers, M. (1981) 'Mechanism, Superaddition, and the Proof of God's Existence in Locke's *Essay*', *Philosophical Review*, 90(2), 210–51.

Birch, T. (1756, 1757) *History of the Royal Society*, 4 vols (facsimile 1968).

Boas, M. (1949) 'Hero's *Pneumatica*', *Isis*, 40, 38–48.

Boas, M. (1952) 'The Establishment of the Mechanical Philosophy', *Osiris*, 10; reprinted New York, 1981.

Boylan, M. (1980) 'Henry More's Space and the Spirit of Nature', *Journal of the History of Philosophy*, 18(4), 395–405.

Boyle, R. (1772) *Works*, 6 vols.

Brinton, C. (1950) *Ideas and Men.*

Burnham, F. B. (1974) 'The More–Vaughan Controversy: The Revolt against Philosophical Enthusiasm', *Journal of the History of Ideas*, 35, 33–49

Burtt, A. E. (1932) *The Metaphysical Foundations of Modern Physical Science*, 2nd edn.

Cajori, F., ed. (1946) *Sir Isaac Newton's Mathematical Principles ... Trans. by Andrew Motte in 1729*, 2nd edn, Berkeley

Campagnac, E. T. (1901) *The Cambridge Platonists, being Selections from the Writings of Benjamin Whichcote, John Smith and Nathanael Culverwel*, Oxford.

Bibliography

Casini, P. (1984) 'Newton: The Classical Scholia', *History of Science*, 22(1), 1–58.

Cassirer, E. (1953) *The Platonic Renaissance in England*.

Christie, R. C. ed. (1886) *The Diary and Correspondence of John Worthington*, vol. II(2), Chetham Society, vol. 114.

Cohen, I. B. (1958) *Isaac Newton's Papers & Letters on Natural Philosophy*, Cambridge, Mass.

Cohen, I. B. (1964) ' "Quantum in se est": Newton's Concept of Inertia in relation to Descartes and Lucretius', *Notes and Records of the Royal Society*, 19(2), 131–55.

Cohen, I. B. and A. Koyré (1961) 'The Case of the Missing *Tanquam*: Leibniz, Newton and Clarke', *Isis*, 52, 555–66.

Colie, R. L. (1957) *Light and Enlightenment: A Study of the Cambridge Platonists and the Dutch Arminians*, Cambridge.

Cope, J. I. (1956) *Joseph Glanvill, Christian Apologist*, St Louis.

Copenhauer, B. P. (1980), 'Jewish Theologies of Space in the Scientific Revolution: Henry More, Joseph Raphson, Isaac Newton and their Predecessors', *Annals of Science*, 37, 489–548.

Coudert, Alison (1975) 'A Cambridge Platonist's Kabbalist Nightmare', *Journal of the History of Ideas*, 36, 633–52.

Cudworth, R. (1845) *True Intellectual System*.

Culverwel, N. (1669) *An Elegant and Learned Discourse of the Light of Nature with several other Treatises*, Oxford.

Curtis, M. (1959). *Oxford and Cambridge in Transition, 1558–1642*, Oxford.

Debus, A. G. (1978) *Man and Nature in the Renaissance*, Cambridge.

Descartes, R. (1974) *Oeuvres . . . Nouvelle Presentation*, ed. B. Rochot et al., vol. 5, Paris.

Dobbs, B. J. T. (1975) *The Foundations of Newton's Alchemy*, Cambridge.

Feingold, M. (1984) *The Mathematicians' Apprenticeship*, Cambridge.

Feisenberger, H. A., ed. (1975) *Sale Catalogues of Eminent Persons*, vol. II, *Scientists*.

Field, J. V. (1988) *Kepler's Geometrical Cosmology*.

Fraser, A. (1984) *The Weaker Vessel: Woman's Lot in Seventeenth Century England*.

Fulton, J. F. (1961) *Bibliography of Robert Boyle*, 2nd edn, Oxford.

Gabbey, A. (1974) 'Avertissement' in Descartes (1974) vol. 5, pp. 628–47.

Gabbey, A. (1977) 'Anne Conway et Henry More: Lettres sur Descartes (1650–51)', *Archives de Philosophie*, 40, 379–404.

Gabbey, A. (1982) 'Philosophia Cartesiana Triumphata: Henry More (1646–1671)' in T. M. Lennon, J. M. Nicholas and J. W. Davis (eds), *Problems of Cartesianism*, Kingston and Montreal, pp. 171–250.

Gale, Th. (1669–77) *The Court of the Gentiles*.

Gibbon, E. (1910) *Decline and Fall of the Roman Empire*.

Glanvill, J. (1661) *The Vanity of Dogmatizing*.

Glanvill, J. (1666) *Philosophical Considerations touching Witches and Witchcraft*.

Glanvill, J. (1671) *A Praefatory Answer to Mr. Henry Stubbe*.

Glanvill, J. (1676) *Essays on several Important Subjects in Philosophy and Religion*.

Glanvill, J. (1682) *Lux Orientalis*, 2nd edn.

Glanvill, J. (1688) *Saducismus Triumphatus*, enlarged edn.

Greene, R. A. (1962) 'Henry More and Robert Boyle: On the Spirit of Nature', *Journal of the History of Ideas*, 23, 451–74.

Grene, M. and D. Nails (1986) *Spinoza and the Sciences*, Dordrecht, Boston Studies in the Philosophy of Science, vol. 91.

Guerlac, H. (1963a) *Newton et Epicure, Conférence au Palais de la Découverte, 2 Mars 1963*, Paris (see also Guerlac 1977: 82–106).

Guerlac, H. (1963b) 'Francis Hauksbee: Experimentateur au profit de Newton', *Archives Internationales d'Histoire des Sciences*, 16, 1113–28 (see also Guerlac 1977: 107–19).

Guerlac, H. (1967) 'Newton's Optical Aether', *Notes and Records of the Royal Society*, 22, 45–57 (see also Guerlac 1977: 120–30).

Guerlac, H. (1977) *Essays and Papers in the History of Modern Science*, Baltimore.

Gunther, R. T. (1931) *Early Science in Oxford*, VIII, Oxford.

Hall, A. Rupert (1948) 'Sir Isaac Newton's Notebook, 1661–65', *Cambridge Historical Journal*, 9(2), 239–50.

Hall, A. Rupert (1980) *Philosophers at War: The Quarrel between Newton and Leibniz*, Cambridge.

Hall, A. R. and M. B. Hall (1962) *Unpublished Scientific Papers of Isaac Newton*, Cambridge.

Hall, A. R. and M. B. Hall (1964) 'Philosophy and Natural Philosophy: Boyle and Spinoza', in I. B. Cohen and R. Taton (eds), *Mélanges Alexandre Koyré*, Paris, 241–56.

Hall, A. R. and M. B. Hall (1969, 1971, 1975, 1977) *The Correspondence of Henry Oldenburg*, vols VII, 1969; VIII, 1971; X, 1975; XI, 1977, Madison, Milwaukee and London.

Hall, Marie Boas, ed. (1971) *The Pneumatics of Hero of Alexandria. A facsimile of the 1851 Woodcroft Edition*.

Hall, Marie Boas (1987), 'Koyré and the Development of Empiricism in the later Renaissance', *History and Technology*, 4, 225–33.

Harrison, J. (1978) *The Library of Isaac Newton*, Cambridge.

Henry, J. (1987) 'Medicine and Pneumatology: Henry More, Richard Baxter, and Francis Glisson's *Treatise on the energetic nature of substance*', *Medical History*, 31(1), 15–40.

Henry, J. (1990) 'Henry More *versus* Robert Boyle: The Spirit of Nature and the Nature of Providence', in Hutton 1990.

Hesse, M. B. (1961) *Forces and Fields: The Concept of Action at a Distance in the History of Physics*.

Hill, C. (1965) *Intellectual Origins of the English Revolution*, Oxford.

Hiscock, W. G. (1937) *David Gregory, Isaac Newton and their Circle*, Oxford.

Hunter, M. (1982) *The Royal Society and its Fellows, 1660–1700*, Chalfont St Giles, British Society for the History of Science, Monographs no. 4.

Hutton, S. (1984) 'Reason and Revelation in the Cambridge Platonists and their Reception of Spinoza' in K. Grunder and W. Schmidt-Biggemann (eds), *Spinoza in der Frühzeit seiner Religiösen Wirkung*, Heidelberg.

Hutton, S., ed. (1990) *Henry More (1614–87) Tercentenary Studies*.

Jacob, A. (1985) 'Henry More's *Psychodia Platonica* and its relationship to Marsilio Ficino's *Theologia Platonica*', *Journal of the History of Ideas*, 46, 503–22.

Jacob, M. C. (1976) *The Newtonians and the English Revolution, 1689–1720*, Hassocks.

Jentsch, H. G. (1935) *Henry More in Cambridge*, Göttingen.

Keill, John (1726) *An Introduction to Natural Philosophy or Philosophical Lectures read in the University of Oxford Anno Dom. 1700*, 2nd edn.

Kepler, J. (1966) *The Six-cornered Snowflake* (1611), trans. C. Hardie, Oxford.

BIBLIOGRAPHY

Kirkby, J. (1734) *Mathematical Lectures. By Isaac Barrow.*

Klibansky, R. (1939) *The Continuity of the Platonic Tradition during the Middle Ages.*

Koyré, A. (1957) *From the Closed World to the Infinite Universe,* Baltimore.

Koyré, A. (1968) *Metaphysics and Measurement,* ed. M. A. Hoskin, Cambridge, Mass.

Kristeller, P. O. (1979) *Renaissance Thought and its Sources,* New York.

Kuhn, T. S. (1977) *The Essential Tension,* Chicago.

Lenoble, R. (1943) *Mersenne ou la Naissance du Mecanisme,* Paris.

Leyden, W. von (1968) *Seventeenth Century Metaphysics.*

Lichtenstein, A. (1962) *Henry More: The Rational Theology of a Cambridge Platonist,* Cambridge, Mass.

Lindberg, D. C. and R. L. Numbers (1986) *God and Nature: Historical Essays on the Encounter between Christianity and Science,* Berkeley.

Locke, J. (1976–) *Correspondence,* ed. E. S. de Beer, Oxford.

Lovejoy, A. O. (1948) *The Great Chain of Being,* Cambridge, Mass.

McGuire, J. E. (1977) 'Neoplatonism and Active Principles: Newton and the *Corpus Hermeticum*', in R. S. Westman and J. E. McGuire, *Hermeticism and the Scientific Revolution,* Los Angeles.

McGuire, J. E. and P. M. Rattansi, (1966) 'Newton and the Pipes of Pan', *Notes and Records of the Royal Society,* 21(2), 108–43.

McGuire, J. E. and M. Tamny, (1983) *Certain Philosophical Questions: Newton's Trinity Notebook,* Cambridge.

MacKinnon, F. I. (1925) *The Philosophical Writings of Henry More,* New York.

McMullin, E. (1978) *Newton on Matter and Activity,* Notre Dame and London

Mamiani, M. (1979) *Teorie delle Spazio da Descartes a Newton,* Milano.

Marion, H. (1880) *Franciscus Glissonius quid de natura substantiae seu vita naturae senserit . . .,* Paris.

Mazzolini, R. G. and S. A. Roe (1986) *Science against the Unbelievers: the Correspondence of Bonnet and Needham, 1760–1780,* Oxford.

Merton, R. K. (1965) *On the Shoulders of Giants,* New York.

Metzger, H. (1938) *Attraction Universelle et Religion Naturelle chez quelques Commentateurs anglais de Newton,* Paris.

Mintz, S. I. (1962) *The Hunting of Leviathan,* Cambridge.

More, H. (1642) *Psychodia Platonica, or a platonicall song of the soul,* Cambridge.

More, H. (1646) *Democritus Platonissans; or, An Essay upon the Infinity of Worlds out of Platonick Principles,* Cambridge.

More, H. (1647) *Philosophical Poems,* Cambridge.

More, H. (1650) *Observations upon Anthroposophia Theomagica and Anima Magica Abscondita by Eugenius Philalethes* [Thomas Vaughan] *by Alazonomastix Philalethes.*

More, H. (1653a) *An Antidote against Atheisme, Or an Appeal to the Natural Faculties of the Minde of Man, whether there be not a God* [reprinted in CSPW, 1662, 1712].

More, H. (1653b) *Conjectura Cabbalistica: Or a Conjectural Essay of interpreting the minde of Moses, according to the threefold Cabbala* [reprinted in CSPW].

More, H. (1656) *Enthusiasmus Triumphatus: Or a Discourse of the Nature, Causes, Kinds, and Cure of Enthusiasme,* Cambridge [reprinted in CSPW].

More, H. (1659) *The Immortality of the Soul, So farre forth as it is demonstrable from the Knowledge of Nature and the Light of Reason*, Cambridge [reprinted in CSPW].

More, H. (1660) *An Explanation of the Grand Mystery of Godliness: Or a True and Faithful Representation of the Everlasting Gospel*, London and Cambridge.

More, H. (1662) *A Collection of Several Philosophical Writings of Dr. Henry More [CSPW]*, London and Cambridge. Also prints for the first time More's letters to Descartes followed by the next item:

More, H. (1664a) *Epistola H. Mori ad V.C. quae Apologiam complectitur pro Cartesio* [separate issue].

More, H. (1664b) *A Modest Enquiry into the Mystery of Iniquity*. Followed by *Synopsis Prophetica: or, The Second Part of the Enquiry* and *The Apology of Dr Henry More*, Cambridge.

More, H. (1667) *Enchiridion Ethicum*, London and Cambridge.

More, H. (1668) *Divine Dialogues*/two volumes: 'The First Three Dialogues', 'The Last Two Dialogues'.

More, H. (1671) *Enchiridion Metaphysicum*, London and Cambridge.

More, H. (1676) *Remarks upon two late ingenious Discourses; the one, an Essay, touching the Gravitation and Non-gravitation of Fluid Bodies; the other, touching the Torricellian Experiment.*

More, H. (1675–79) *Opera omnia*.

More, H. (1712) *CSPW, The Fourth Edition*.

Newton, I. (1687) *Philosophiae naturalis Principia mathematica*.

Newton, I. (1704) *Opticks, or a Treatise of the Reflections, Refractions, Inflections & Colours of Light*.

Newton, I. (1706) *Optice*.

Newton, I. (1713) *Philosophiae naturalis Principia mathematica*, Cambridge.

Newton, I. (1952) *Opticks*, New York.

Newton, I. (1959–77) *Correspondence*, ed. H. W. Turnbull et al., 7 vols, Cambridge.

Nicolson, M. H. (1929a) 'Christ's College and the Latitude-Men', *Modern Philology*, 27, 35–53.

Nicolson, M. H. (1929b) 'The Early Stages of Cartesianism in England', *Studies in Philology*, 26, 356–74.

Nicolson, M. H. (1930) *The Conway Letters*, New Haven.

North, J. D. and J. J. Roche (1985) *The Light of Nature: Essays in the History and Philosophy of Science presented to A.C. Crombie*, Dordrecht.

Oldenburg, H., ed. (1671) *Philosophical Transactions*, 72 (19 June), 2182–4.

Oldenburg, H., ed. (1676) *Philosophical Transactions*, 122 (21 February), 550.

Osler, M. J. and P. L. Farben, eds (1985) *Religion, Science and World-View: Essays in Honour of Richard S. Westfall*, Cambridge.

Passmore, J. A. (1951) *Ralph Cudworth: An Interpretation*, Cambridge.

Patrides, C. A. and R. B. Waddington, eds (1980) *The Age of Milton: Backgrounds to Seventeenth Century Literature*, Manchester.

Peile, J. (1900) *Christ's College*.

Peile, J. (1910) *Biographical Register of Christ's College 1505–1905*, Cambridge.

Pepys, S. (1974) *Diary*, ed. R. Latham.

BIBLIOGRAPHY

Power, J. E. (1970) 'Henry More and Isaac Newton on Absolute Space', *Journal of the History of Ideas*, 31, 289–96.

Powicke, F. J. (1926) *The Cambridge Platonists: A Study*.

Purver, M. (1967) *The Royal Society: Concept and Creation*.

Randall, J. H. Jnr (1940) 'The Development of Scientific Methodology in the School of Padua', *Journal of the History of Ideas*, 1, 177–206.

Raven, C. E. (1950) *John Ray, Naturalist*, Cambridge.

Raven, C. E. (1953) *Natural Religion and Christian Theology. The Gifford Lectures, 1951, First Series*, Cambridge.

Ray, J. (1713) *Three Physico-Theological Discourses*.

Robinson, H. W. and W. Adams (1935) *Diary of Robert Hooke*.

Roe, S. A. (1981) *Matter, Life and Generation: eighteenth century embryology and the Haller-Wolff debate*, Cambridge.

Roger, J. (1963) *Les Sciences de la Vie dans la Pensée française du XVIIIe siècle*, Paris.

Rogers, G. A. J. (1985) 'Descartes and the English' in North and Roche (1985), 281–302.

Rossi, Paolo (1968) *Francis Bacon: From Magic to Science*.

Rudrum, A. (1984) *The Works of Thomas Vaughan*, Oxford.

St Augustine (1968) *The City of God against the Pagans*, trans. D. S. Wiesen.

Sarton, G. (1953) *A History of Science*, I, Oxford.

Saveson, J. E. (1960) 'Differing Reactions to Descartes among the Cambridge Platonists', *Journal of the History of Ideas*, 21, 560–7.

Schmitt, C. B. (1966) 'Perennial Philosophy: Steuco to Leibniz', *Journal of the History of Ideas*, 27, 505–32.

Shapin, S. and S. Schaffer (1985) *Leviathan and the Airpump*, Princeton.

Shapiro, B. J. (1969) *John Wilkins, 1614–72*, Berkeley.

Shorthouse, J. (1881) *John Inglesant*.

Shugg, Wallace, et al. (1972) 'Henry More's *Circulatio Sanguinis*', *Bull. Hist. Medicine*, 46, 180–9.

Spingarn, J. E. (1920) *A History of Literary Criticism in the Renaissance*, New York.

Stahl, W. H. trans. (1952) *Macrobius, Commentary on the Dream of Scipio*, New York.

Staudenbaur, C. A. (1968) 'Galileo, Ficino and Henry More's *Psychathanasia*', *Journal of the History of Ideas*, 29, 565–78.

Strong, E. W. (1934) *Procedures and Metaphysics*.

Strong, E. W. (1970) 'Barrow and Newton', *Journal of the History of Philosophy*, 8/2, 155–72.

Syfret, R. H. (1950) 'Some Early Critics of the Royal Society', *Notes and Records of the Royal Society*, 8/1, 20–64.

Taylor, T. trans. (1818) *Iamblichus' Life of Pythagoras*.

Thackray, A. (1970) *Atoms and Powers*, Cambridge, Mass.

Thomas, K. (1971) *Religion and the Decline of Magic*.

Thorndike, L. (1941) *History of Magic and Experimental Science*, V, New York.

Toulmin, S. E. (1959) 'Criticism in the History of Science: Newton on Absolute Space, Time and Motion', *Philosophical Review*, 68, 1–29, 203–27.

Tulloch, J. (1874) *Rational Theology and Christian Philosophy in England in the Seventeenth Century*, vol. II, *The Cambridge Platonists*, Edinburgh and London.

BIBLIOGRAPHY

Turnor, E. (1806) *Collections for the History of the Town and Soke of Grantham. Containing authentic Memoirs of Sir Isaac Newton.*

Van Helden, A. (1985) *Measuring the Universe*, Chicago.

Walker, D. P. (1958) *Spiritual and Demonic Magic from Ficino to Campanella.*

Walker, D. P. (1972) *The Ancient Theology: Studies of Christian Platonism.*

Walker, D. P. (1986) *Il concetto di spirito o animo in Henry More e Ralph Cudworth*, Naples.

Ward, R. (1710) *The Life of the Learned and Pious Dr. Henry More.* There is an unprinted continuation of this book in the Library of Christ's College, Cambridge, pressmark Dd.7.12.

Webster, C. (1969) 'Henry More and Descartes: Some New Sources', *British Journal for the History of Science*, 4(4), 359–77.

Webster, C. (1975) *The Great Instauration.*

Werkmeister, W. E. ed. (1963) *Facets of the Renaissance*, New York.

Westfall, R. S. (1962) 'The Foundations of Newton's Philosophy of Nature', *British Journal for the History of Science*, 1, 171–82.

Westfall, R. S. (1971) *Force in Newton's Physics.*

Westfall, R. S. (1972) 'Newton and the Hermetic Tradition', in A. G. Debus (ed.), *Science, Medicine and Society in the Renaissance*, II, 183–98.

Westfall, R. S. (1975) 'The Role of Alchemy in Newton's Career' in M. L. Righini Bonelli and W. R. Shea (eds), *Reason, Experiment and Mysticism in the Scientific Revolution*, New York, 189–232.

Westfall, R. S. (1980) *Never at Rest: A Biography of Isaac Newton*, Cambridge.

Whewell, W. (1860) *Mathematical Works of Isaac Barrow*, 2vols., Cambridge.

Wiener, P. P. and A. Noland, eds (1957) *Roots of Scientific Thought*, New York.

Willey, B. (1949) *The Seventeenth Century Background.*

Wood, A. (1817) *Athenae Oxoniensis*, Oxford.

Yates, F. A. (1964) *Giordano Bruno and the Hermetic Tradition.*

Index

Académie Royale des Sciences, 195
Aglaophemus, 27
Agrippa, Henry Cornelius, 35
alchemy, 37–8, 98–100, 107, 232–4
Alfred, King, 22
Alleyn, Giles, 260
Ambrose, St, 110
Amyclas, 111
Anaxagoras, 76
Anaximander, 27
Apollonios, 53
Apuleius of Madoura, 21, 24
Aquinas, St Thomas, 30, 31
Archimedes, 7, 53, 55, 183, 190, 252
Archytas, 15
Aristobulus, 110
Aristotle, 14, 32, 42, 47, 53–4, 68, 77, 83, 116, 146–7, 164, 176, 199
Artapanus, 30
Ashmole, Elias, 277
Aston, Francis, 169
Atlas (astrologer), 27

atomism, *see* Epicureanism, mechanical philosophy, *and names*
Augustine, St, 20–2, 24–5, 27–8, 30, 48, 61
Averroes, 42
Avicenna, 42

Bacon, Francis, 33–4, 49–50, 54, 179–80, 218
Bacon, Roger, 252
Bainbridge, Thomas, 88
Baines, Sir Thomas, 94, 101–2, 132, 168
Barrow, Isaac, 169, 203–4, 209–13, 216, 222–3, 255, 270, 277
Basil, St, 20
Basil, Valentine, 36
Baxter, Richard, 127
Beale, John, 166, 188
Beaumont, Joseph, 89
Benedetti, G. B., 54
Bentley, Richard, 228, 237–8, 246, 270
Bessarion, Cardinal, 26